Mining in the Americas

HELMUT WASZKIS

Mining in the Americas

Stories and history

Best wishes
Helmut W

WOODHEAD PUBLISHING LIMITED
CAMBRIDGE ENGLAND

Published by Woodhead Publishing Ltd,
Abington Hall, Abington,
Cambridge CB1 6AH, England

First published 1993

British Library Cataloguing in Publication Data
A catalogue record for this book is available from the British Library.

ISBN 1 85573 131 2

Designed by Geoff Green (text) and Chris Feely (jacket)
Typeset by Best-set Typesetter Ltd., Hong Kong
Printed by St Edmundsbury Press, Suffolk, England

For my son PETER MARC as he
follows in the footsteps of his father
and his grandfather into the exciting
world of mining and metals and on to
the discovery that there is more to
metals than trading . . .

Contents

Preface

Jacob Fugger the Rich (1449–1525) was most probably the world's first global metal trader. He was also a multinational merchant banker and a maker of kings. It was Fugger money that financed the wars of the Spanish crown, and it was Fugger money that bought the imperial crown – of the Holy Roman Empire of the German Nation – for Charles I of Spain, to make him Emperor Charles V in 1517.

Jacob Fugger was also a mining tycoon with mines and smelters all over Europe. It was he who called mining 'God's Greatest Gift' or, in his language, the old German, '*Groest Gotsgab*' – and the Spanish galleons soon brought gold and silver, the first mining wealth of the New World, by the shipload to Europe.

These are stories and histories about metal mining in the western hemisphere. The theft of myriad pre-Columbian golden adornments and artefacts by the Spanish conquistadors, the feverish pursuit of placer gold and the subsequent continent-wide discoveries of innumerable mines of ever more types of metals – all this activity spans 500 years, and has never before been brought together in one volume. An attempt to do so is made here.

The reader who looks for lusty stories, adventure, lucky finds and tragedies, as well as an attempt to view mining in a general economic context, will not be disappointed. A reader who, however, expects a learned treatise on ore bodies, reserves and statistics, will look for them in vain. This book was intended to relate only stories.

The subject is vast, the setting large; the author hopes he has been able to do justice to the subject.

No more opportune time for the publication of this volume than the 500th anniversary of Columbus' landing in the Caribbean. What Columbus 'discovered' was only discovery to European eyes. He and those who followed, the conquistadors, the conquerors, destroyed entire civilizations and decimated the native populations, before they built again. The flow of metals from the Americas changed Europe, and contributed to the dawn of the modern age.

The metric system has been used throughout. Tonnes equal 1000 kilograms. Precious metal contents as well as the value of old coins have been converted into kilograms and tonnes. And when reference is made to production of copper, or similar, the basis is always the metal content.

Helmut Waszkis
April 1993

Acknowledgements

The material contained in this book was first presented to the public in a series of lectures at the Geosciences Graduate School of the State University of Campinas, UNICAMP, in Campinas, state of São Paulo, Brazil, between August and October 1991. The author sincerely thanks the professors and graduate students of the Geosciences Institute for their support, encouragement and critical appraisal.

A book such as *Mining in the Americas* could not possibly have been written without the generous support of many people. I have received such support, and much encouragement, during the almost 10 years that this book has been in the making. I am deeply grateful to all who responded with great enthusiasm to my requests for help. These acknowledgements can in no way reward these people adequately nor reflect my gratefulness. The following list is only a partial listing; it follows no particular order or system.

- Katherine Dorr Abreu and Maria Bernadette Amâncio de Sá Alves of IBRAM, the former Instituto Brasileiro de Mineraçao in Belo Horizonte, for information on CVRD, iron ore in Brazil and the 'Iron Square in Minas Gerais'.
- Diana Kinch, the Latin American editor of *Metal Bulletin*, who was most generous, and unstinting, in her support and encouragement, and who pointed out many errors and gave much up-to-date information on the Brazilian situation. She even agreed to review the final typescript and eliminated quite a few German expressions that had slipped into the English text.

- Professor Dr Luiz Augusto Milani, Professor Dr Celso Pinto Ferraz, Professor Dr Saul B Suslick and many others of the Instituto Geociencias of Universidade Estadual de Campinas, Campinas, SP for their support and encouragement, advice and many discussions on specific points in the book.
- Dr Cassio de Souza Melo and Eric de Champeaux de la Boulaye of Rio de Janeiro and Porto Velho, Rondônia, who introduced me to the Amazon jungle of Brazil, to tin mining, and to the gold *garimpos* of Rondônia.
- Wolfgang Oskar Mayer of Konstanz, Germany, a former colleague, and friend for three decades, who knows Peru and Bolivia, was familiar with the mining industries of these countries, and was able to give many valuable suggestions, hints on errors, and general observations.
- Rolf Kantorowicz of Cologne, Germany, who lived many years in Bolivia and Peru and whose recollections of events during and shortly after World War II in Bolivia contributed meaningfully to the text.
- Dr Elisa Carrillo and Sr Kathleen Connell of Marymount College, Tarrytown, NY, accompanied my somewhat belated return to undergraduate college studies after my retirement from business, and encouraged my renewed pursuit of historical research and critical appraisal of 'things of the past'.
- Oscar Burchard of the Ore & Chemical Corporation, New York, also a colleague many years ago, and a friend of three decades, in many countries and different businesses.
- Henry Rothschild of Hartsdale, NY, for information on the events surrounding the nationalization, by Bolivia, of Mina Matilde, and for critical comments regarding the role of the Spaniards in the conquest of South America.
- Guenther Vollmers and Vilma Salazar of Lima, Peru, Hans Tordoir of La Paz, Bolivia, Jorge Lazo of Santiago de Chile, and Victor Vazquez of Mexico, DF, for assisting with innumerable requests for books, statistical data and for clarification of many points.

Statistical consistency and accuracy are essential for a book like *Mining in the Americas*. I have tried to adhere to this principle. Practically all production figures were taken from the excellent annual statistical tables published by Metallgesellschaft AG, Frankfurt am Main. Dr W Hergenhans gave me all possible support. Special thanks go to him and his colleagues.

It was only for gold that I relied on the excellent publications of Goldfields. And the data for Mexico came from the Secretaria de Energia, Minas y Industria, Paraestatal.

I alone am responsible for the opinions and conclusions expressed here.

Introduction

Mining, metal-making and the use of metals have been of primary importance in the development of man and of civilization. Next to the cultivation and preparation of food, man's growing dominance and application of metals permitted progress. And as people became more and more the masters of metals, of the art and craft of making them and putting them to ever more varied and more sophisticated uses, they developed technology and industry. The processes were mutually inspiring: better metals brought about more uses; more uses required more metals; more demand caused more search; prospecting advanced the technology of mining and brought new discoveries.

Gold may well have been the first metal that caught man's eye: it was colourful, shiny and, in the beginning, easily available when present in rivers and creeks. The glittering pebbles could easily be picked up. And the metal was malleable. Beating it with stones allowed easy shaping. It took little technology and effort to convert golden pebbles or nuggets into sheet, and into shapes for jewellery and decoration.

Copper was probably the second of the metals primitive people found. This one they saw in veins and lumps, but in the earth and in rocks, not washed clean in rivers. And where they found it, it probably came in larger quantities than alluvial gold nuggets in riverbeds.

Man probably next discovered that copper was harder than gold and that it could be used for tools. Further steps along the long road towards the metal ages were the – no doubt also accidental – discoveries that the more one hammered copper the harder it became, and that fire – very intense, hot fire – eventually made metal liquid and helped to purify it. Accidental discoveries no doubt also were that certain coloured and colourful dirts, or rocks, separated into metal and slag when heated sufficiently. And finally that mixing one metallic substance, like copper, with another, like tin, through heating and liquefaction, gave another metal, bronze, that had yet other qualities. Similarly, copper could be mixed with zinc to give brass.

We do not know when man first found gold but we can safely assume that the date must have been very old, in the dawn of man's beginnings. Copper was probably found and used much later. Still later man found silver, lead and then the other metals.

As people's awareness grew that some of the metallic substances could be shaped into tools and weapons, and the demand for the metals became greater, their search for supplies led to new sources and eventually to mining.

Mining and the making of metals accompany closely the technological progress of man. Without metals, most of man's technology is unthinkable. And it was the discovery of ever more metals, different ones, as well as different alloys, that eventually led to today's very sophisticated technologies.

The Americas – North, Central and South – are today integrated parts of the world, of the global economy, and of the intricate flow of metallic resources and finished products. The nickel of Canada, the tin, iron and bauxite of Brazil, the tin of Bolivia, the lead, zinc and copper of Peru, and the copper of Chile – to mention only very few – are important parts of the worldwide supply of these metals and of the economies of the countries under review.

But this was not always so . . .

And it all really began with gold . . .

CHAPTER 1

Pre-Columbian origins

Metal mining in the Americas did not begin with Columbus and the Spaniards. It had, in fact, begun much earlier. How much earlier, we do not know. But we should not doubt that some form of metal gathering and processing commenced in the second millennium before Christ.

Gold adornments, jewellery and artefacts were offered to the Spaniards by the natives of the Caribbean islands. The visitors bartered for more. But soon they simply took what gold they saw. It was no different in Mexico and Central America, Colombia and Venezuela, and then in Peru, Bolivia and Chile as well. The museums of these countries, as well as many in Europe and the United States fortunately own quite a few pieces that have survived.

Many of these adornments and artefacts reveal high degrees of craftsmanship, great artistic expression and considerable sensitivity. Most of the pieces have come from graves that survived the centuries untouched. Still today, new pieces are occasionally offered by illegal dealers who get them from grave robbers. Rarer still, but nevertheless happening from time to time, are discoveries of entire untouched tombs. Such finds then allow studies by professionals that enrich our knowledge of the pre-Columbian peoples.[1]

But highly developed though their artistic ability was, they were not accomplished miners. What gold the pre-Columbians used was wash-gold from rivers and creeks as well as dry riverbeds. It was all alluvial gold, generally of high purity, and in the form of nuggets and dust. The nuggets were usually pounded into sheet and then moulded. But the pre-Columbians also had solid pieces, clearly the result of melting down and pouring into moulds. Silver and copper were probably taken from readily available veins that surfaced in many

The first meeting between the Europeans and the native Americans.

places.[2] Copper was apparently found in varying degrees of purity, some mixed with tin and other impurities that made for harder metals. The pre-Columbians thus had what probably were accidental alloys of considerable hardness which allowed them to shape tools that could be used as adzes, knives and axes, including what they called 'tumis' that were even used for rather delicate surgery like trepanation.[3]

While most of the pre-Columbian mining was thus the collecting or recovery of surface metal occurrences, some kind of primitive sluicing may have existed on the Andean slopes, in Mexico and even in the Caribbean. And so probably did some primitive hacking into rock. Where particularly rich veins surfaced, it is logical to assume – and the chroniclers make mention of such cases – that the pre-Columbians had battered the rock, had broken in, and had opened short adits to get at the metal by following the veins. There is evidence of pre-Spanish copper mining in Chuquicamata, Chile; the Paracas civilization of Peru, from around the time of Christ, has yielded metal fishing hooks.[4] It was apparently the Incas who introduced bronze.[5] The famous cinnabar mine of Santa Barbara, near Huancavelica, Peru was, when first shown to the Spaniards, a colourful outcropping that the Incas had exploited superficially and on a very small scale to produce colouring powder – a rouging agent for their women.[6]

But it must be admitted that we know very little about mining in pre-Columbian America. Do we know enough to conclude that what there was was technically primitive and on a very small scale? Such judgement takes nothing away from the high degree of artistic achievement and beauty of what the early peoples of the Americas produced. Unfortunately, the greater part that fell into the hands of the Spaniards was destroyed. Few pieces of the original finds survived; most of what we admire today has come to the museums in the last 50 years.

However, while the pre-Columbians probably knew little of mining, they did have considerable skill in metallurgy. The early Peruvians, the early Colombians (from the Caucá valley in today's Colombia) as well as the early Mexicans knew melting and even smelting. They also knew and applied casting, soldering, repoussé and gilding.[7] Bronze and brass alloys no doubt came about accidentally rather than by design, and they did not know iron.[8] But they may have known how to reduce cinnabar ore to mercury, and they may even have developed quite on their own a primitive amalgamation process.[9]

While pre-Columbian gold and silver artefacts were found all over the Americas, the most elaborate and artistically most intricate were really those from Mexico, Colombia and Peru.

In Mexico, the first meaningful 'finds' were what the Spaniards plundered from the Aztecs. But most of the pieces were apparently not of Aztec origin. What the Aztecs had were gifts, or tribute, or plunder they had collected from other tribes they had subjected.

In Colombia, the gold finds are generally attributed to the Chibcha, but we know too little to say categorically that these were the artists.

In Peru, however, we know that almost nothing of what the Spaniards obtained was of Inca workmanship. Some was of Moche, or Mochica origin but most came from the Chimu empire which had extended along most of Peru's northern coast and was said to have been as wealthy as that of the Incas in the mountains, until the Incas conquered and subjugated it – taking with them to Cuzco not only many Chimu gold treasures but also many Chimu artists.[10]

When we look today at examples from all three countries, we have difficulty in stating categorically which of these pre-Columbian civilizations produced the most beautiful artefacts and adornments. And there is probably no need to decide – all three produced fabulous pieces.

On the other hand, what adornments and artefacts the Spaniards found on the Caribbean islands were generally of little beauty. It was to be no different on the northern continent, once they landed in Florida and further north, nor along the coast of Brazil, nor when the Spaniards and the Germans first penetrated Venezuela and Colombia.

We should now turn to the Caribbean, where it all began . . .

Notes

1. *National Geographic Magazine* of Washington, DC had extensive articles in its issues of October 1988 and June 1990 about recent discoveries of unspoiled tombs in the Chimu/Moche area of northern Peru.
2. Purser, *Metal Mining in Peru*, page 23. Purser's book is still today the single most important review of mining in Peru and, with its introductory chapters, footnotes and annexes, of special value to anyone who wishes to study the mining scene of Peru and western South America and the interrelationships between the present and the past. Purser's dispassionate analyses are exemplary.
3. Tumis are found mainly in the Andean countries. They varied in size and shape from 1 inch instruments to tools the size of a hatchet.
4. Purser, page 21.
5. Ibid, page 22.
6. Several of the old chroniclers mentioned this.
7. Purser, page 22.
8. Ibid.
9. This is my own theory. Several of the old chroniclers mention that some primitive furnaces were found and it is quite conceivable that the Incas or their predecessors in the area discovered the roasting process and that the fumes quickly precipitated again and formed small balls of 'liquid metal'.
10. The Incas did produce quite a number of gold artefacts. Much of what the first Spaniards saw by way of life-size animals in the golden garden in Cuzco seem to have been pieces of solid gold, and because of a certain roughness of design were probably of Inca origin – a classic simplicity of line and form as opposed to the more intricate, baroquish forms and designs of the Moche artefacts.

    Pre-Columbian Peruvian gold artefacts actually came from locations now called Chavin, Piura, Vicus, Batan Grande, Chanchan, Chancay, Paracas, Nazca and others. But the most expressive pieces were of Mochica origin (about AD 100–500) from the Piura area, and of Chimu origin (AD 800–1300) from Chanchan, near today's Trujillo.

CHAPTER 2

Early Spanish endeavours

The pursuit of wealth – greed – was clearly the principal motivation that drove Christopher Columbus and the Spaniards to sail west. To prove that the earth was round, to satisfy yearnings for adventure and discovery, to take the faith to other lands, and even the search for spices, were of secondary importance. Columbus' diary mentions the word gold 114 times. God is only mentioned 26 times, Christ only 3 times.

Why this hunger for gold?

By about the year 1500, Europe had pretty much run out of money. Its stock of coins in circulation was pitifully small. Barter, the generalized means of exchange in the Middle Ages, was now too limiting, too cumbersome and hence unsatisfactory for long distance business – with the then existing means of transportation. And the mines of Central Europe were running out of silver and gold. Much of the existing stock of coins had flown east, to Asia, in exchange for the silks and spices of the Orient.[1]

The reader should be aware that the term 'mining wealth' is used in what follows somewhat loosely. Wealth means riches, material possessions, as well as objects that have both utility and exchange value. 'Commodities of great value' could be another term. What the first European conquistadors found were worked objects of value such as idols, adornments, and jewellery. But they also obtained dust and nuggets. Most – practically everything – was quickly melted into bars, chains and other shapes, thus becoming simple metallic commodities of no utility. It continued that way throughout the centuries. Gold and silver moved in the form of bullion, copper in bars, other metallic substances in bars or concentrates and often even simply as ores.[2]

Mining has probably been the most important industry of the western hemisphere during the last 500 years, ever since Columbus reached the Caribbean in 1492. Yet while the industry has contributed greatly to the wealth and development of some countries, like the United States, Canada, Brazil, Chile and Mexico, it has had much less apparent effect on the development of others, such as Peru and Bolivia.[3] Or, if indeed there was an effect, at some point or period throughout the centuries, it was of brief duration and soon dissipated. And while the mining wealth flowing from the Americas to Spain in the sixteenth and seventeenth centuries was instrumental in making Spain a world power for a while, it also almost immediately contributed to weaken the very same empire and to hasten its downfall.[4]

The most lasting result of the flow of the mining wealth – gold and principally silver – from the Americas to Europe in the sixteenth and seventeenth centuries was probably that it stimulated commerce in Europe, and contributed to the manufacture of consumer goods on an ever-expanding scale which in time led to the industrialization of Europe.[5]

Hispaniola

Christopher Columbus and his men reached the western hemisphere on 12 October 1492. They made landfall at one of the islands of the Bahamas group. Several of the natives they met wore golden adornments and insistent enquiries brought more. But the natives and their huts were primitive, no organized society seemed to exist, and the Spaniards found neither meaningful quantities of gold nor any mines. And when they pointed to the pieces of gold and to the land, the natives seemed to gesticulate south, to other islands.

The Spaniards sailed on, west, then south, then west again, and then south again. By 27 October, they reached what the natives apparently called Cubanacan, which Columbus called Juana and which we know as Cuba. But though they sailed up and down the coast and even rowed up a few rivers, the dense vegetation and the complete absence of gold discouraged them.

They sailed on once more, now eastward, beyond the eastern tip of Cuba and then along the northern shore of Marien, or Bohio, which later came to be called Española, or Hispaniola and which we know today as Haiti and the Dominican Republic. There Columbus founded Navidad, the first European settlement in the New World, near modern-day Cap Haïtien, in Haiti. It was Christmas Day 1492. Columbus' flagship Santa Maria had run on to a reef. The cargo was saved and the ship materials went into the building of a fort.

And here now was gold – in the form of dust and nuggets, and also as adornments. The nearby rivers were apparently full of gold. It had only to be picked out of the gravel.[6]

Soon Columbus was on his way back to Spain; he had spent no more than three weeks exploring the northern shore of the island. By the end of March 1493, he was again in Seville, to report, to gain glory, and to prepare a much larger venture. The cost of the first voyage seems to have been something like 2 million maravedis; what he brought back from the New World was worth but a few thousand maravedis. But there were also a few precious stones, exotic birds, plants, fruits and even a few natives.[7]

By September 1493, barely six months later, Columbus was off once more, and now with a large fleet.

It does well to pause briefly. Where the first voyage, the one of 1492, saw some 90 people in all set out to search for the Indies, and in three small vessels, the second voyage, a mere one year later, saw 17 ships and over 1200 people! What had happened? Suddenly they knew that 'there', 'out there', was indeed a 'New World' and wealth, and everybody wanted a share.

Such a large throng and so much enthusiasm can, however, hardly have been caused by the few handfuls of gold Columbus had brought back from the first voyage. It must have been really much more the stories he and his men told, how they had heard of 'more gold further on', to the west, to the south, of 'an island entirely made of gold'. The island was never found, but the stories spread ever wider and further, and with each retelling they became more fantastic.

For the second trip, Columbus received very specific instructions from the crown, including rules on how to organize and account for the collection of gold.[8]

He now took a more southerly route, made landfall in the lesser Antilles, and then worked his way west, discovering Dominica, Guadaloupe, Montserrat, Antigua, Jamaica and Boriquén – which the Spaniards named San Juan and which later became Puerto Rico – before he reached Hispaniola and the first settlement.

But Navidad had meanwhile been wiped out. Greed, for gold and women, had led to disputes with the natives and fatal encounters. Columbus sailed on, back east. Near modern-day Puerto Plata they founded a new settlement and called it La Isabela. Rich gold deposits said to be nearby made the choice of location easy.[9]

We may safely assume that among the large throng of people Columbus took with him on his second trip were several who had notions of mining, even if professional miners were not as yet part of the group.[10]

Small parties were sent inland to explore. These did find gold, in several rivers and even encrusted in rocks. One party advanced as far as Cotui, or Cotuy, which in time was to become a major mining centre, principally for its copper deposits.

More gold adornments and artefacts were found, or bartered for with the local inhabitants, or simply taken, during the second, third and fourth voyages of Columbus. These trips also intensified contact with Puerto Rico, Jamaica,

Venezuela, Honduras, Nicaragua, Costa Rica and Panama, to add to the Bahamas, Cuba and Hispaniola discovered during the first voyage.

Without much delay, Columbus decided to send his second in command back to Spain, with 13 vessels and the gold so far collected – 11 000 maravedis.[11] The crown's contribution to the cost of the voyage had amounted to 3 million maravedis.[12] The pickings were indeed small so far.

Early in 1494, the crown was petitioned to send gold washers and miners from Almaden, to wash, or pan, river sand as well as to dig into the ground.[13] And when, also during the second voyage, Columbus landed briefly in Panama, at a place called Veragua, and traces of gold were discovered, it was the first time the metal was found on the continent, on 'Tierra Firme'.[14]

Columbus himself sailed around, exploring first the coast of Cuba and then discovering Jamaica. On his way back he sailed along the southern coast of Hispaniola, past the mouth of the Ozama river where a few years later, in 1497, his brother Bartolomew Columbus would found the city of Santo Domingo.[15]

It was during these two years, 1494–6, that the island of Hispaniola was systematically searched,[16] river by river and creek by creek. The explorers dug holes and maybe even a few adits. Gold was all over, always placer gold in riverbeds. Many of the Indian chiefs they saw wore golden adornments.

On this, his second trip, Columbus stayed well over two years, from October 1493 to February 1496. Sporadic hostile encounters with the natives soon became all-out war.

The single-minded greed of the Spaniards was coupled with total disregard for the natives. These were considered, at least at the beginning, heathen without rights and without possessions. They could thus be used, and without compensation. As there were so many of them, the death of a few mattered little. They were simply replaced.[17]

The Spaniards were not humane employers; they were also neither good organizers nor knowledgeable miners. Theirs was a warrior tradition, schooled as they were in the battles against the Moors. Many of them were the impoverished sons of estate owners or of civil servants. The natives, in turn, had never worked with discipline. They had never produced anything beyond their daily needs. To now pick nuggets and pan gold dust out of river beds was beyond their comprehension. Their food supplies included no reserves,[18] and work days of 10 to 12 hours, or even more, did not allow for food production. The Spanish law, or the general rules, seemed to require that the Spaniards had to supply minimum subsistence foods. But in the beginning, the Spaniards had no food at all, other than what they could get, or take, from the natives.

The system was slavery at its worst, with disastrous results. Where the number of inhabitants on Hispaniola at the time of the arrival of the Spaniards in 1492 was estimated at 200 to 300 000, by 1508 only some 60 000 were still counted.[19] And no more than 12 000 remained by the year 1517.[20] A major smallpox epidemic, in 1519, decimated the population even further.

Nobody had, of course, counted heads in 1493, and by 1517 more than 12 000 must have been alive but hidden in the hills; still, whatever the numbers, brutal decimation, almost extermination, it no doubt was.

The pattern was similar elsewhere and at about the same time – in Cuba, Puerto Rico and the other Caribbean islands. And it was soon to be repeated on the mainland. That the methods employed were no different from those used elsewhere, in Africa and even in Europe (we only have to read about the atrocities during the Thirty Years Wars in Central Europe, in the early seventeenth century), that human lives counted for little in those times, and that people were insensitive to the sufferings of others, are hardly acceptable excuses.

Mining in the western hemisphere had started off on the wrong foot. But gold it certainly produced.

Complete statistics about how much gold the Spaniards got from Hispaniola are not available. We have already mentioned what small quantities were shipped with the first two sailings back to Spain, in January 1493 and January 1494. We also know that by 1519 the island was practically mined out as by then production was down to almost zero.[21]

Already in 1497, the crown had ordered Columbus to set up at La Isabela a mint for coins[22] but it was not until 1544[23] that a royal mint was established. It began to mint copper coins, quartos and medio quartos. The copper came from the copper mine at Cotui. They also coined 1, 2, 3, 4 and 8 real pieces for which they used impure silver imported from Peru.

By 1496, Columbus had overcome most of the native resistance. He then again returned to Spain, to render accounts and to prepare yet a third voyage during which he was to discover Guyana, Venezuela and the island of Margarita from which would come substantial quantities of pearls. During his absence, so-called technicians in the search for gold arrived, most probably German miners.[24]

Late in the year 1500, a court lawyer by the name of Francisco de Bobadilla arrived in Hispaniola as its new governor. Columbus had asked for a lawyer and judge to help put order into the affairs of the New World. Bobadilla instead sent Columbus back to Spain in chains.[25]

Bobadilla lowered the tax on gold from one-third to one-tenth[26] which at once made him immensely popular. He also paid a number of claims on the crown – with the confiscated assets of the Columbus brothers.[27]

But he only ruled for two years, 1500 to 1502, and was followed by Nicolás Ovando from 1502 to 1509, and then by Diego Columbus, the admiral's son, from 1509 to 1515. By the time Diego Columbus was recalled, the native population had practically disappeared, and so had the gold.

During the Ovando years, a total of 2 million ducats of gold (about 7.6

Metal-working in the sixteenth century: hand-minting coins by hammering slugs of metal into moulds (although this scene is from Germany, we know that the first mints in the Americas used precisely the same methods and tools). Source: Woodcut by O. Magnus, seventeenth century.

tonnes) were apparently produced, or about 300 000 ducats a year.[28] By 1517, the quantity was down to 120 000 pesos (about half a tonne).[29]

Hispaniola probably had its apex of gold production during the Bobadilla and Ovando years, essentially the first decade of the sixteenth century. Not that these two were the best, or better governors; it was simply that prior to Bobadilla the island had been systematically searched, and likely riverbeds and creeks had been identified. Also, enough natives were still around to serve as labour. And, even as the number of natives began to diminish, slaving trips to nearby islands, mainly the Bahamas but also the Lesser Antilles,[30] as well as to Tierra Firme – which meant Yucatán, Central America and the shores of today's Venezuela and Colombia – replenished the numbers.

The gold bearing zone of Hispaniola was roughly along a wide sash of land between today's Santo Domingo, Puerto Real on the north coast and as far west as Lares de Guahaba in today's Haiti.[31]

During the Ovando years, two meltings (of gold) took place each year at each of two foundries, or smelteries, at Buenaventura and La Vega. The sources say that '500 000 pesos were obtained'.[32] Still, we know that about 2 million ducats were produced during the years 1503 to 1509. May we assume, or extrapolate, that a similar quantity had been produced during the preceding 10 years, and that yet another such quantity was produced during the following 10 years, until 1520? This would give us a total of 6 million pesos for the first 27 years of Spanish occupation of Hispaniola. Another similar amount must have been stolen – or shall we say, euphemistically, 'produced and not reported to the crown'?

A grand total of 12 million ducats produced during the period 1493 to 1520 would give us, at 3.75 grams per ducat, a grand total of about 45 tonnes of gold. (As we shall see later, 19 tonnes is what the Customs House in Seville registered as officially entered between 1503 and 1530 from all of the 'Indies'.)

Forty-five tonnes of gold over a period of 27 years would come to about 1.7 tonnes per year. By comparison, in 1990, the Dominican Republic produced about 4 tonnes of gold.[33]

It was, of course, all dust and bullion that they shipped out in those old days – but also here nothing has changed. It is still bullion, and often only doré, a silver/gold alloy with a few impurities.

The concept of 'theft' needs a few words of explanation. Non-reporting, or smuggling, or theft of gold and silver production, be it mining production or the finding, or plunder, of artefacts and jewellery, plagued the Spaniards from the first day – and it has never ended. It continued throughout all the years of the Spanish colonial rule (and Portuguese rule in Brazil) and then, later on, throughout the republican years, well into our times. Estimates of the percentages that were not reported vary but many chroniclers and historians agree that it may well have been as much as 50%. Of the emeralds of Colombia, an even higher percentage was probably smuggled out. And this undoubtedly also applied to the diamonds of Brazil. In recent years, substantial quantities of Brazilian placer gold – panned and washed in the Amazon region, often by uncontrolled and uncontrollable prospectors – appeared as exported from Uruguay, a country never known for a single gold mine.

By 1517, Hispaniola was already growing sugar cane and making sugar. In fact, the first sugar had been produced as far back as 1506, and within a few years, sugar was to replace gold as the principal industry and export item. The terrible smallpox epidemic of 1519 had reduced the native population to about 4000, probably barely enough for household servants and a few panning operations.[34]

With no more cheap native labour and no more easy (and rich) gold finds and deposits, gold mining came to a virtual standstill and ceased to exist as a meaningful industry as from about the year 1520.[35]

The export statistics show a curious pattern. Whereas in about 1500 gold represented 100% of exports, and still dominated by 1520 with about 90%, it was down to 33% by 1530 and ceased to be listed by 1540. By 1560, sugar represented 66% and hides 33%. By the year 1600, hides had become the principal export item, with 55%, while ginger was second with 11%. Hides then grew in importance and accounted for 65% by the 1650s. And by 1680, hides had become the only export item. By 1740, cattle represented 90% and hides 10%, and by 1810, lumber or logs accounted for 50%, tobacco for 35% and hides for the rest.

Gold does not reappear in any statistic of the colonial period. Nor did miners appear in an island-wide census taken in 1606, not even in Cotui.[36] But then, the entire seventeenth century was considered 'lost'. Smuggling and depredations by pirates had caused the crown and the local authorities to lose control over great parts of the northern coast. It was then decided to simply 'devastate' the northern shore area by destroying plantations and buildings and withdrawing people to the south, around Santo Domingo. This was done in about 1605/6. As a consequence, Hispaniola withdrew from the world; for a while maritime contact with the rest of the world practically ceased.

Still, some gold continued to be found, or produced, but it was not picked up statistically; it was used as countervalue for the purchase of smuggled goods – those brought in by pirates and itinerant merchants. Some panning no doubt continued to take place in selected – and usually secluded – locations across the island, throughout the centuries. But we should also be aware just how severe the fluctuations of fate and fortune were.

The export statistics given above need some interpretation. While ginger was still the product of organized labour, the production of hides and eventually simply cattle was much less so. Cattle had been introduced quite early, and apparently ran wild and proliferated. As gold panning and then sugar production declined, and no other meaningful industry replaced them, people left the island. Most of the Spaniards that had come with high hopes and in great numbers in the first 30 years after the conquest now left, disappointed and often destitute. And then the pirates, or buccaneers, added insecurity. Hides were eventually produced simply by shooting cattle that ran wild.

The population curve, based on what we should call 'more-or-less accurate population counts' shows the approximate figures given in Table 2.1.[37]

The economy of Hispaniola picked up again early in the eighteenth century. Between 1650 and 1690, French interests began to occupy the western part of

Table 2.1. The population of Hispaniola, 1500–1800

Year	Inhabitants
1500	125 000
1700	15 000
1800	110 000

the island which, as from 1697, became Haiti, a French colony, and which then saw massive imports of African slaves and a rapidly expanding and soon booming cane sugar industry.

The island then went through major changes, with successive and extremely violent revolutions in Haiti, French and then Haitian occupation of the eastern part, and the eventual independence movement which resulted in the Dominican Republic.[38] Subsequently, the island went through more convulsions and more stagnation.

It is really only since about the middle of the twentieth century that the Dominican Republic has reached some kind of equilibrium in its development that seems to augur reasonably well for the future. But still, they first had to get rid of the Trujillo dictatorship which had strangled the country for 31 years (1930–61). The turmoil in Haiti persists.

Mining continued, and still does. At the midpoint of the twentieth century it was still a minor sector, amounting in 1960–70 to about 1.5% of gross national product (GNP). But it picked up in 1973 when it jumped to 5% of GNP. What made the difference was ferronickel, which as from 1972 began to be exported in sizeable quantities.[39] The statistics of the early 1980s show exports of ferronickel, gold, silver and bauxite. The quantities of precious metals were significant (see Table 2.2). The bauxite deposits are apparently small and the quantities exported minor and rather sporadic. Still, the exports of the mining sector amounted to about 25 to 30% of the country's total exports of goods – in terms of value and not counting income from the tourist

Table 2.2. Exports of gold and silver, 1981–3 (tonnes)

	1981	1982	1983
Gold	12	12	11
Silver	62	67	41

Table 2.3. Exports from the mining sector, 1987–9 ($ million)

	1987	1988	1989
Ferronickel	115.2	308.8	372.0
Bauxite	1.3	2.7	—
Doré (gold/silver bullion)	120.0	98.1	88.8
	236.5	409.6	460.8
Total exports	711.3	890.1	927.7
Mining products as % of total exports	33	46	50

industry.[40] And the picture had not changed much by the late 1980s, as can be seen in Table 2.3.[41]

The export of these mining products brings the country more dollars than any other manufacturing industry. More dollars flow into the country only from services like tourism and money transfers from Dominican citizens working abroad.

The history of mining in Hispaniola ends thus today on an upbeat note, despite all the stories of tragedies and plunder and decline. The island is, admittedly, very poor and suffers principally from overpopulation and insufficient industry. It is, however, noteworthy that mining does again play a role, and now a major one in the development of the country.

Cuba

Columbus discovered Cuba in 1492, during his first trip, and some superficial exploration took place. But the quantities of gold found were negligible and as a consequence further exploration and settlement were delayed. In fact, when the native population of Hispaniola began to diminish drastically, as from about 1506, slaving expeditions to neighbouring islands including Cuba were authorized. In 1511, when the then viceroy Diego Columbus dispatched Diego Velazquez to conquer Cuba, Velazquez was accompanied by one Hernán Cortés who acted as his treasurer. On the island they encountered little resistance and Velazquez was soon appointed governor.

Some gold was panned out of the rivers and small parties began to penetrate the interior. In 1530, rich copper veins were discovered. But the discoverers did not know what to do with the ore. By 1543, a German specialist in smelting copper ore, one Hans Tetzel from Nuernberg, arrived. For two years he tried unsuccessfully to process the ore, but it was only after his return to Germany that he managed to solve the problem. In 1545, the Spanish authorities gave him a 10 year exclusive contract, whereupon Tetzel returned to Cuba to set up a smelter near Santiago.[42] Tetzel produced metal until his death in 1571, after which members of his family continued the operation.

Some of the copper was used locally to produced copper kettles, or large pots, for the sugar mills. Part of the production was also exported, principally, no doubt, to Hispaniola, but also to other islands of the Caribbean area. But it did not amount to much. Though a number of small mines and smelters continued to operate, the mining industry went through the seventeenth century in a general state of stagnation.[43] It was the government which in about 1700 gave another push to reactivate the industry. But the crown had difficulty in finding operators willing to lease mines. It is probably impossible today to evaluate clearly what happened. Maybe the deposits were too low grade, or the technology had been lost; the fact is that copper metal was imported from Peru and Mexico.[44]

It was only from about 1830, after Spain's monopoly on investment in and trade to and from the colonies had been broken, that foreign interests started to take a look at Cuba. It was an English company which began mining the Santiago del Prado mine[45] (already mentioned in 1705[46] and which may well have been the mine that Tetzel had worked in the sixteenth century) and from which, by 1838, some 10 000 tonnes of ore had been exported.[47]

Several other foreign companies also began operating about the same time but were apparently not very successful. Still, it would be daring to speak of a mining industry in Cuba in those years. The statistics for the period 1833–7 show that metals or metallic products accounted for 10% of all imports while precious metals accounted for 6.25% of all exports – a sure sign that some gold was still being mined.[48] By 1854, some 31 mining companies were operating around Santiago de Cuba. Most were apparently of US and British origin; they employed 2389 people, 1300 of them slaves. It is not quite clear what these mines produced, apparently mostly handpicked, high grade copper ore. One firm, Compañia San José, had a furnace where a 70% copper-containing metal was produced.[49]

The Cubans claim they produced in those years, 1830–40, about 12% of the world's copper, and that Cuba was the principal supplier of copper to the United Kingdom.[50]

The industry's fundamental problems were fuel and transport. The absence of coal, and by then also the absence of wood and charcoal, were major handicaps. By the early 1830s, many regions of Cuba were already totally deforested.[51] And transport was handled with mules and camels.[52] By 1843, a railroad had been built from the mines to Santiago, but lower copper prices after the American Civil War and bitter disputes between the railroad and the principal mine, La Consolidada, led to the closing of the latter by 1868.[53]

Different mines and numbers appear in the old documents, such as a mine called San Augustin, in Bayatabo, that produced 50 tonnes in 1851, another called Buena Esperanza that produced 75 tonnes in 1852, as well as one near Cienfuegos, the name of which we do not have, that shipped about 780 tonnes in 1856 of a grade of 17% copper.[54]

Gold continued to be produced at a place called Auras,[55] and there was knowledge of a chrome ore deposit and of quite a few iron ore deposits.[56]

In 1868, the independence movement which had already changed South America between 1808 and 1826 (and had generally ruined the economies of most of the countries), also brought wars and confusion to Cuba. One war was to last 10 years, until about 1878, only to lead to another, in 1895, which was to last another three. The wars completely paralyzed all mining activities.[57] It was only in 1903 that La Consolidada, now in US hands, began operating once again.[58] Earlier already, in 1858, the manganese deposits of the Oriente province had begun production, but it was only in 1902/3 that sizeable quantities were exported. Also the iron ore mines were opened and shipments began, 21 000 tonnes in 1884. The Spanish-American Iron Company exported 74 000 tonnes in 1895; by 1902/3, exports had risen to 350 000 tonnes.[59]

The war for independence (1895–6) was followed by US occupation of Cuba (1898–1902) which, though it favoured US owned mines, left the rest of the country stagnating once more.

There is not much to write about mining in Cuba in the twentieth century. Very little material is available, and what there is gives scarce information. We are obviously handicapped by the dearth of information coming out of Cuba under the Castro government, in power since 1959, for the current situation as well as for the recent past.

We do know that the mining industry picked up again after 1899,[60] that Cuba had unfavourable trade balances in 1899, 1900 and 1901,[61] but that the shortfall was more than made up by foreign investment, principally in mining. But we do not know in which mines and in what ores.

A reciprocal trade agreement with the United States, signed in 1902, mentioned copper ores and concentrates, chrome ores, copper residues and iron ores. These could enter the United States duty-free.

US investments in the island grew, from $45 million in 1896 to $205 million in 1911 and to $1.1 billion in 1927. Over half of this was in the sugar industry and a quarter in railroads and government bonds. An unspecified part of the investment was in mining – but again we do not have details.

By 1940, US investment was down to $560 million, a drastic decline. Apparently, government bonds had been repaid and part of the investment in the sugar industry had been repatriated.[62]

Whatever mining there was in Cuba cannot have amounted to much as sugar exports in the 1930s accounted for between 75 and 82% of total exports. Tobacco must have accounted for a major part of the remainder.[63]

By the early 1990s, Cuba was a substantial producer of nickel, from a number of mines and three plants located in the north-easternmost part of the island. The reserves are estimated to be the second largest in the world, and once a fourth processing plant can be financed, Cuba may also become the second largest producer of nickel in the world, behind Canada. But, for the time being, these are still dreams. In 1989, the last year for which detailed statistics are available, nickel production reached 46 500 tonnes (contained nickel), but thereafter production apparently again tapered off, to about 42 000 tonnes in 1990 and about 34 000 tonnes in 1991. In 1992, it was expected to once more reach 40 000 tonnes.

Cuba also produces about 2000 tonnes of copper per year and a similar quantity of cobalt. Chrome ore production fluctuates in the range of 30–50 000 tonnes per year. Raw steel production stood at 314 000 tonnes in 1989.

And gold? It is obviously there, and is no doubt being produced, but precise data are not available.

Puerto Rico

The Spaniards landed in Puerto Rico in November of 1493, during Columbus' second voyage. The natives called their island Boriquén; the Spaniards baptized it San Juan and then called it San Juan Bautista. It was only after 1521 that it came to be called Puerto Rico.

It was not until 1508 that the Spaniards began settlement. It was not that the island was left alone for 15 years. It was, after all, not far from Hispaniola, and it was from Hispaniola that slaving expeditions began in 1506. But no doubt even earlier, small parties of explorers must have searched the island for gold.

In 1508, rumours of large quantities of gold started a stampede.[64] Many of those who were dissatisfied in Hispaniola because of finding little gold, or because of being excluded by the then existing administrative group that ran the island,[65] emigrated. But in Puerto Rico it was not much different. There the settlers again found only placer gold, in rivers and creeks, and only in relatively small quantities. Again they enslaved the natives, worked them to death, and got relatively little gold.

And, again, as in Hispaniola, the natives resisted the encroachment, gave battle, tried to withdraw into the interior and were decimated. A major battle was fought in 1512. The Spaniards used dogs, specially trained to attack natives. Those natives caught were made slaves and were branded with an 'F', apparently to identify them as property of the king, Ferdinand of Aragon.[66]

The decline of the native population followed a similar pattern to that in Hispaniola. Estimates for the population[67] at the time of the arrival of the Spaniards vary from 30 000 to 600 000,[68] whereby the second figure appears absurd, the first rather more realistic. Thirty-six years later, in about 1530, apparently no more than about 2–3000 natives were left.

The reduction of the native population soon led to imports of African slaves, beginning in about 1510.[69]

Already by the late 1520s mining activities began to decline, mainly because not enough gold was found any more. Much of the European population was restless. When, in about 1532/3, several ships came in search of horses for the conquest of Peru, and the crew members told of the wealth of that country, many of the unemployed and idle joined them and left Pureto Rico. The governor had to take severe measures to avoid wholesale departure.[70]

Blacks substituted the natives as mine workers. They supposedly brought both smallpox and bubonic fever, but because they were used to these diseases they suffered less; it was the natives that suffered most.[71] But the crown limited the influx of blacks to avoid too lopsided a racial profile.[72] As a consequence, not enough hands were around and still more Spaniards decided to leave.

There is not much more to write about metal mining in Puerto Rico. Some gold panning probably continued throughout the centuries but the deposits were minor and the extraction minute. In time deposits of copper, molybdenum, and chrome ore became known, but they were not of meaningful commercial size.[73]

Jamaica

The island of Jamaica, discovered by Columbus during his second voyage, also served as a slaving ground as the Spaniards explored and exploited the gold placers on Hispaniola, in the years 1500 to 1520. In about 1518, the first placers in Jamaica itself were apparently discovered. By then, the local population had already been pretty much decimated. Not much gold was found, however.

In time, Jamaica was to become a haven for an ever larger population of pirates and thus a collection point for much of the gold and silver loot the buccaneers took from the Spanish galleons and towns.

It was to be only in the twentieth century that Jamaica became a major mining country when its bauxite mines were opened and the island became a prime supplier of raw materials to the world's aluminium producers.

In 1991, Jamaica was among the three largest bauxite producers in the world, with 11 609 000 tonnes. But it must be noted that production fluctuated widely in previous years, with a low of 7.4 million tonnes in 1988.

In alumina Jamaica is also among the three largest producers in the world, with 3 015 000 tonnes in 1991; but here also the production rhythm varied widely in prior years.

The Danes and the Brandenburgers

Here is a short story about greed and the lure of gold in the Caribbean islands.

In about 1650, the Danes, and a few years later also the Germans, felt the urge to seek gold. The Danes formed a company, the Danish West Indies Co., to go after black slaves in Africa and to trade with the buccaneers. In 1666, they set sail for the islands and took possession of St Thomas. The first ship brought 63 settlers straight from the Copenhagen jails. They found only three natives. Two fled, the third had one leg chopped off to prevent him from also running away.[74]

It is a lusty tale, of contraband, jailbirds, killings, confusion, wealth and poverty – those first 10 years of the Danes on St Thomas. In about 1675, the Brandenburg elector, Friedrich Wilhelm of the Hohenzollern family, also succumbed to gold fever and, in around 1683, two Prussian expeditions left Europe, one to set up a slaving outpost in Africa, the other for St Thomas to

Table 2.4. Estimates of early gold extraction

Years	Tonnes
1503–5	5
1511–20	9
1521–30	5

establish a partnership with the Danes, the bases for which had apparently been agreed upon between Copenhagen and Berlin: a rare sight – Germans as junior partners of Danes. Naturally, they did not hit it off. The Danes, the senior partners and rulers of the island, demanded payment of taxes; the Germans resisted. The Danes next seized the warehouses of the Germans; confusion reigned between Copenhagen and Berlin. But in 1701, when the Brandenburg elector became king of Prussia, the joint venture with the Danes was liquidated. The Germans withdrew.

The Danes held on to the island until 1916 when they sold it to the United States. But neither gold nor any other metals were ever found on the island, neither by the Danes, nor by the Brandenburgers, nor by the Americans.

Estimates of early gold extraction

There are some estimates of what quantities of gold were extracted, or panned, in the Caribbean islands during the first years of the conquest. The sources for such calculations are the records of the Customs House in Seville through which, at least in theory, all imports from (and exports to) the Americas had to flow.[75] See Table 2.4. As much was probably stolen, we can safely assume that real production must have been at least double, if not triple or quadruple the quantities that were registered.

Meanwhile . . .

As these early endeavours of the Spaniards were taking place in the Caribbean, John Cabot, an Italian, reached Labrador, Newfoundland, and Nova Scotia in 1497/8, while in 1500 the Portuguese explorers Vincente Yáñez Pinzón and Pedro Álvarez Cabral reached Brazil.

Right after Juan Ponce de Leon discovered Florida in 1512, Vasco Nuñez de Balboa reached the shore of the Pacific in Panama in 1513, and Juan Diaz Soliz sailed up the River Plate in 1516. Simultaneously, the Caribbean area was being crisscrossed by thousands of explorers, adventurers and conquistadors. News of the existence of a rich empire – the Aztec – beyond Yucatán, and of another – the Inca – south of Panama, along the Pacific coast, had reached even Spain and led increasing numbers of adventurers to flock to the New World.

Mining for gold had already begun. We do not know when the first professional miners crossed the Atlantic, but by 1526 the Welsers, merchant bankers of Augsburg, Germany, the second most important financial and trading house (next to the Fuggers) of the Spanish realm, had made arrangements with the Spanish crown that gave them privileges and preferences, among them the right to take German miners – in those years the most skilful in Europe – to Cuba. Similar arrangements undoubtedly also existed for Hispaniola where they even set up a *factoria*, a kind of office/warehouse/administrative headquarters. And when the Welsers obtained from the crown exclusive rights to explore and exploit Venezuela – which quickly led to the pursuit of El Dorado ('the golden one') – more German miners must have crossed the ocean.

But we must first turn to Mexico . . .

Notes

1. Moya Pons, *Historia Colonial de Santo Domingo*, page 51; also Cassá, *Historia Social o Economica de la Republic Dominicana*, page 27; see also Braudel, *The Mediterrarean and the Mediterrarean World in the Age of Philip II*, volume 1, pages 464ff.
2. And essentially out of South and Central America and even out of Canada.
3. It can be argued that the railroads of both countries only became economically feasible once they were in place. See also Chapter 10.
4. Cassá, page 16.
5. Several authors deal with this subject, especially Braudel. Also see Moya Pons, page 244. He says that 'by the end of the 17th century, the greater part of the ships that plied the transatlantic trade were foreign built and 5/6th of the trade with the Indies was conducted by foreigners'.
6. Rodriguez Martinez, *Panoramica de la Isla Hispaniola, Siglos XI–XVI*, page 41. Here appears the word dust, which can only be the result of panning.
7. The amount given for the cost of the voyage is small but comes from the original documents. Two million maravedis (the then most common Spanish coin), at 450 maravedis for 1 peso de oro gives us, expressed in gold, about 20 kilograms of gold (a peso de oro weighed 4.601 grams). But Columbus did not buy the ships, he only rented, or chartered them.
8. The haste with which the second trip was put together, and the number of people demanding to be included in the group, allows us to conclude that professional mining men, who probably would have had to be brought from Germany or at least from the Almaden mine, were not as yet included.
9. The *New York Times* of 27 November 1990 published, under the heading 'Dominican Bluff Yields Columbus's First Colony', a detailed report by John Noble Wilford about recent excavations near the town of El Castillo, in the Dominican Republic. The site is apparently La Isabela.

> Digging at the storehouse site, archaeologists accidentally released thousands of evanescent drops of mercury that glittered in sunlight as they floated away. With the look of a triumphant alchemist on his face, Dr José M Cruxent, a Venezuelan archaeologist, cupped a brownish glob of thickly liquid mercury in his hand and uttered the intoxicating word, 'gold'. This was some of the mercury the Spaniards had brought for separating from the mineral matrix the pure gold they had hoped to find in abundance.

If true, the find and the conclusions from it would be amazing. After all, La Isabela was founded in 1493 and was supposedly abandoned only five years later. None of the old documents mentioned that mercury was taken to the Americas that early.

10. Rodriguez Martinez, page 54. He does mention one expert in mining, a Fermin Cade.
11. Ibid, page 64. A maravedi was a gold coin in use under the Moors in southern Spain, and probably went back as far as the eleventh century. Soon after the conquest of the Caribbean islands by the Spaniards, maravedis are mentioned and are referred to as 'minor copper coins'; also, but later, as made of vuiton, a copper alloy. I have not been able to determine whether the old records refer to values expressed in old gold maravedis, or in copper coins. It would appear odd that values for the cost of Columbus' trip, as well as the first returns in gold, were expressed in low value copper coins, when soon thereafter, meaning no more than 10 years later, reports mention gold ducats, gold pesos and silver pesos. On the other hand, Columbus set sail with little more than three tiny fishing boats, in today's terms, with no more than a handful of people and few supplies. And he did not buy the ships; he only chartered them.

    I will now include a few basic comments on old coins and values. There is much confusion about the coins and weights that applied at the time of the conquest. The Spaniards brought few coins with them. So they used values expressed in weights.

    Gold was measured in the gold peso (*peso de oro*), also called *castellano*; it was one-fiftieth of a mark – which was an old and rather general European monetary unit, in Spain called *marco*. It weighed about 8 ounces. There were 2 marks to the pound, and a pound was 460.1 grams, a mark 230 grams. One gold peso was thus one-hundredth of a pound, or 4.601 grams. A gold ducat, a coin abolished in Spain in 1530, was five-sixths of a *peso de oro*, or 3.834 grams.

    Silver was then considered (in about 1520) worth one-tenth of gold. A mark of silver weighed 230.1465 grams. There were 67 reales to a mark, or 3.4335 grams to the reale, and then 8 reales to a silver peso (*peso de plata*), or 27.468 grams for a silver peso, also called a *peso de ocho reales*, or a 'piece of eight'.

    Both Purser, *Metal Mining in Peru*, and Hemming, *The Conquest of the Incas*, have excellent summaries on the old coins. Both agree on the weight but Hemming insists a reale weighed 0.6 grams.

    There was also a *peso ensayado*, which was worth 13.25 reales. And, to bring us back to the maravedis, 450 copper maravedis made up one *peso ensayado*.
12. Rodriguez Martinez, page 51.
13. Prieto, *A Mineraçao e o Novo Mundo*, page 50, based on Modesto Bargallo, *La*

Mineria y Metalurgica en la América Española durante la Epoca Colonial, Mexico, 1955.

14. Ibid.
15. Rodriguez Martinez, page 67.
16. Ibid, page 65.
17. Moya Pons, *Historia Colonial*, page 61.
18. Cassá, page 56.
19. Ibid, page 46.
20. Ibid.
21. Ibid, page 65.
22. Rodriguez Martinez, page 129.
23. Ibid, page 131.
24. Ibid, page 74.
25. Ibid, page 81.
26. Cassá, page 40; Moya Pons, *Después de Colón*, page 29.
27. Rodriguez Martinez, page 81.
28. Ibid, page 172.
29. Cassá, page 58.
30. Ibid, page 52.
31. From the map in Cassá, page 42.
32. It is not clear from the old documents whether these figures are 'per year' or refer to the whole period. I choose to be conservative.
33. *Gold*, the journal published by Goldfields Consolidated, 1989, page 18.
34. Which leaves us with the question, 'did the natives die out first so that work could not continue as not enough hands could be found, or did the placers run dry first?' But as gold was again and again found and production today is larger than ever, it was probably the decimation of the natives that reduced production in the late 1510s.
35. Moya Pons, *Después de Colón*, page 167.
36. Ibid, page 96.
37. Cassá, page 135.
38. Independence came late, only in 1844. Though the country ceased to be a Spanish colony in 1795, France, Spain and Haiti and then again Spain battled for control. Full independence was achieved only in 1865.
39. Dauhajre hijo, *Sabado Economico*, page 140.
40. Ibid, page 142.
41. International Monetary Fund (IMF), *International Financial Statistics*, July 1990, page 202.
42. Konetzke, *Fischer Weltgeschichte*, page 300.
43. Le Riverend, pages 68–9. Julio Le Riverend's *Historia Economica de Cuba* is a well-balanced and highly informative book. Though it covers only the period 1697–1951, it helps us to understand the role mining played in the ups and downs of Cuba's fortunes.
44. Ibid, page 69.
45. Ibid, page 201.

46. Ibid, page 69.
47. Ibid, page 201.
48. Ibid, pages 227–8.
49. Ibid, page 369.
50. Ibid.
51. Ibid, page 202.
52. Ibid, page 369.
53. Ibid.
54. Ibid, page 370.
55. Ibid.
56. Ibid.
57. Ibid, page 502.
58. Ibid, page 503.
59. Ibid.
60. Ibid, page 552.
61. Ibid, page 559.
62. Ibid, page 614.
63. Ibid, page 633.
64. Lopes and Figueroa, *O Conocimento Geologico na America Latina*, page 26.
65. Cassá, pages 47–8.
66. Lopes and Figueroa, page 60.
67. Because nobody had counted them.
68. Lopes and Figueroa, page 69.
69. Ibid, pages 79 and 263.
70. Ibid, pages 80–1.
71. Ibid, page 264.
72. Ibid.
73. Ibid, pages 445–6.
74. Arciniegas, *Biografia del Caribe*, page 283.
75. Vilar, *Ouro e Moeda na Historia (1450–1920)*, page 85.

CHAPTER 3

Mexico

The first great bonanza in the history of metal mining in the Americas was the Aztec treasure; it came to the Spaniards as plunder.

One Francisco Hernandez de Córdoba, a wealthy landowner from Cuba, was the first to explore the coast of Mexico, in 1517; with 110 men he landed on the western side of Yucatán. The local Mayas fiercely resisted the intrusion; in a battle near modern Campeche, the Spanish force was cut in half.

The next to try his luck was Juan de Grijalva who, with 240 men, landed at Cozumél in 1518. Again the Mayas resisted; more pitched battles followed. The Spaniards sailed on and eventually reached the area around modern Veracruz where they were finally received with considerable civility – and with impressive gifts of gold.

Both Hernandez and Grijalva, as well as their men, were greatly impressed by the large Mayan cities and the obviously advanced civilization of those they encountered wherever they landed.

No gold was ever found in Yucatán, the heartland of what had been Mayaland. No mines were ever found. What little gold the Mayas had must have been imported from Central America.

The Spaniards hastened back to Cuba, to report. A third expedition sailed the same year, 1518; this one was led by Hernán Cortés. He had with him 508 men. They also made landfall first at Cozumél but soon proceeded to Veracruz.

Moctezuma II, then emperor of Tenochtitlán, as the Aztecs called their country, had heard of the earlier visits by white, bearded strangers, and had thought they were reincarnations of a legendary god, Quetzalcoatl. He tried to placate Cortés, when the latter idled at Veracruz, with golden jewellery, followed by nuggets and dust, and when Cortés sent a message that he and his companions suffered from an illness which only gold could cure,[1] more golden artefacts were sent. This first, substantial 'gift' did not, however, cure the Spaniards. If anything, it only increased their greed for more, made them determined to strike inland and visit the source of this wealth and, in a way, doomed the Aztec civilization.

Among the gifts the Spaniards received were a wheel of solid gold, the size of a Spanish cart wheel, weighing some 3800 gold pesos, and a somewhat smaller wheel of solid silver, weighing 48 marks of silver – about 10 kilograms.[2] They also received much jewellery.

So they set out after the gold. As they proceeded inland, up the sierra, towards what is today Mexico, DF, Moctezuma sent them more presents, hoping to satisfy their greed and asking them not to proceed. But their greed increased.

Once in Tenochtitlán, the Spaniards received yet more gold and silver presents. Their greed was still not satisfied. They enquired where the gold came from and Moctezuma arranged for guides to accompany a few of the Spaniards to look for the mines, most of which turned out to be placer deposits. Moctezuma also delivered the royal treasure – the accumulated

A pre-Columbian goldsmith at work.
Source: the Mendoza Codex.

adornments, jewellery and plates he had in his palaces and temples. Then the Spaniards killed him.

Quite a few of the pieces of this first 'loot' were sent by Cortés to Spain – a first sample of much more to come. But fate, in the person of a pirate, interfered. He, Giovanni da Verrazano, felt he could take from those who had already taken. He was also known as Juan Florentin (because he came from Florence) and even as El Francés, the Frenchman. In 1522, he attacked and plundered the vessel that carried Cortés' first samples of the treasure of Mexico. The loot was taken to La Rochelle and then to the French court – instead of to Spain. Everybody marvelled at the lovely pieces – also at the 150 000 ducats of gold (about 560 kilograms) and 680 lb of pearls.[3] When the Spaniards eventually caught Verrazano, he confessed, before having his throat slit, that he had robbed and sunk some 150 ships.[4]

It is difficult to restrain the pen and not write the adventuresome story of the conquest of Mexico. Few exploits in the history of mankind are equal to the lusty and colourful tales of the bloody conquest of the Caribbean, of Mexico, Peru and Venezuela by the Spanish conquistadors. But the interested reader should look for one of the many specialized histories of these events; this book would have to digress too far, only to discuss superficially ground already covered by others.

We have no reliable means of estimating the value of the precious metals the Spaniards received, and took, in Mexico during the first few years of the conquest. It must have been an orgy of greed. Most of the Spaniards were rough soldiers; the life of a conquistador was not for softies and bookworms; only the toughest, physically and otherwise, survived. And those who survived and succeeded were not exactly used to wealth and easy living, to luxury and good manners. Access to sudden wealth undoubtedly brought other excesses.

We should accept that the treasure they obtained was fabulous. Nothing like it had existed in living memory, and much beyond. No type of superlative expression can possibly exaggerate the beauty, the variety and the value of the treasure.

In 1520, much was lost again when the Aztecs finally rose and forced the Spaniards to leave the city. During the nocturnal flight many of the heavier pieces of jewellery, and most of the bars that had resulted from the melting down of most of the jewellery and artefacts and made to facilitate the division of the loot, were abandoned and thrown off the causeway across which the Spaniards battled their way out.

But giving up the city did not mean abandoning the country. Reinforcements soon strengthened the Spanish forces and the systematic conquest, including the destruction of the old forms of organization and religion, proceeded. It was to be almost three centuries before Mexico threw off the Spanish colonial yoke – and by then it was another Mexico, and another people.

Colonial mining, 1530–1810

Conquest meant also settlement, the introduction of organized mining, and feverish exploration and exploitation.

By 1545, 25 years after the original thrust into Tenochtitlán, major deposits of silver and gold had been identified and were being exploited. In a very few years, production from there exceeded the entire value of the original plunder. But it was as yet only production from finds of almost solid veins of silver. It was to be only in 1555 that the patio process was discovered and applied, as shall be explained below, whereafter lower grade ores could also be processed.

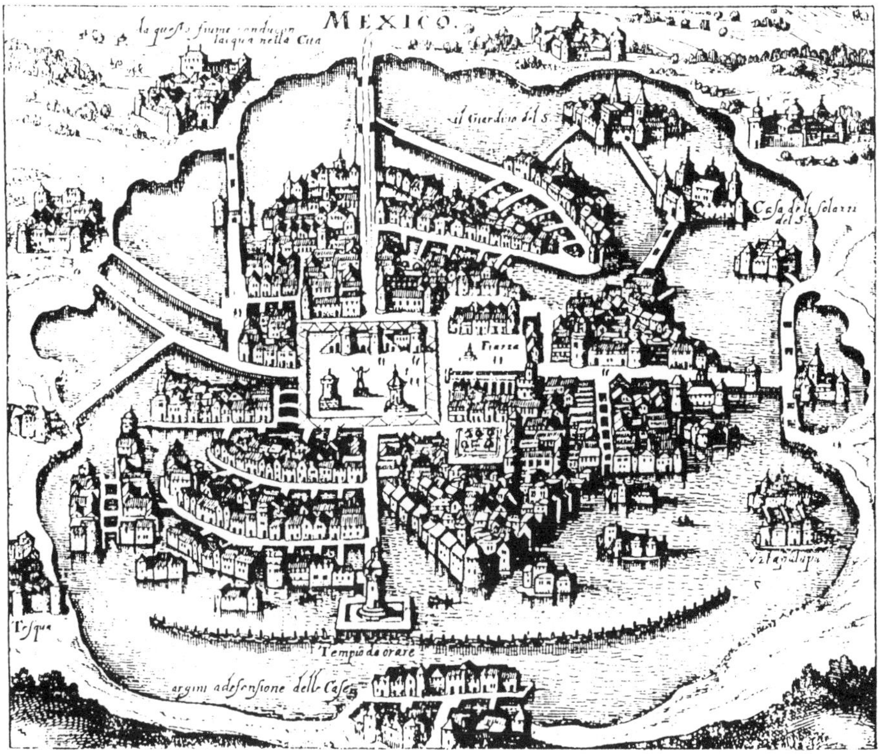

A view of Mexico in about 1593. Source: Johan Ludwig Gottfriedt, *Newe Welt und Americanische Historien*, first published in 1631 by Matthaeus Merian dA and republished in 1980 as Stuttgarter Faksimile Edition im Fackelverlag, Stuttgart, Germany.

The flow of wealth now began. It was soon, following the conquest of Peru in 1533, to become a torrent and, after the discovery of Potosí in 1545, a flood. It not only made Spain wealthy and important – at least for a few years – financing Spain's and the Habsburgs' military adventures in Europe, mainly in Italy. But it also, and most importantly, fuelled Europe's development, industrialization and expansion during the seventeenth and eighteenth centuries. It is not really an exaggeration to say that the pre-eminence in financial, economic and military matters of Europe during those centuries, including worldwide colonization, was brought about by the wealth Europe took out of the Americas – essentially in the form of gold and silver bullion.

Compostela in Nueva Galicia was the first major silver mine struck in 1545.[5] Already earlier, in 1524, small tin and copper deposits had been found in Tasco and, in 1525, silver mines in Jalisco. By 1534, more silver mines were discovered in Tasco. As from 1546, strikes in Zacatecas, Guanajuato, Pachuca, Sombrerete and Fresnillo followed.[6] Another important find, in 1552, was Real del Monte where, in 1555, Bartolomew Medina invented the patio process. These were also the years when iron tools, gunpowder and the use of horses and mules as beasts of burden were introduced.

Technical innovation – amalgamation

The discovery, or rediscovery, of the patio amalgamation process in 1555 by Medina was of great important for the development of gold and silver mining in Mexico, and subsequently in Peru.

The process is quickly explained.[7] Finely ground silver ore was mixed with water, salt, some sulphates and mercury. The resulting muck, or cake, was then stirred for days, by horses, mules or men, walking in it, wading in it, really kneading it. The process at times supposedly took up to 30 days. Water was then used to flush the lighter particles away. The amalgam of mercury and silver, or mercury and gold, remained at the bottom of whatever receptacle they used, was collected, and heated. The mercury then evaporated, and pure, or almost pure silver remained.

This process permitted the working of medium and lower grade silver ores and brought about the second large precious metal boom from Mexico – the first having been the loot of gold and silver adornments. It was to be rather similar in Peru.

It has already been said that there had been in the Americas no real mining – in an industrial sense – before the arrival of the Spaniards. The Americans – the Indians, or pre-Columbian inhabitants, or natives – had at best dug some superficial holes here or there. Even the first Spanish endeavours can have been little more; maybe somewhat deeper holes, or shafts, and some short, shallow adits. The technology to process ores did not exist and thus work could only be pursued where large, wide, rich veins were apparent and justified a major effort.[8]

But the patio process, allowing the processing of lower grade ores, provided the incentive for the search for and exploitation of lower grade material in larger quantities. This need for larger quantities now brought about the need for better mining technology.

This is when mining in the Americas really began, after 1555 in Mexico and after 1571 in Peru. Amalgamation was also the first sophisticated mixing of two or more metallic substances, mercury and silver, also the first industrial separation of a substance into different metals, mercury and silver.

We do not know how soon the Spaniards learned to recover the vaporized mercury by means of simple chimneys for precipitation. But if, as we can assume, the process was not invented in Mexico but was really imported from Europe, the technique to catch mercury for reuse was probably also known.

As regards the origins of the process, Georg Agricola (Bauer), the city physician and mayor of Chemnitz, Germany, who between 1544 and 1556 published his *De Re Metallica*, wrote knowingly and in detail about how to recover gold from ore, and about amalgamation. Although the patio process itself is not mentioned, Agricola gives the formula for the refining of gold – add to 1 lb of metal 1 lb of sulphur, several pounds of salts, 1/3 lb of copper and 1/4 lb of yeast – almost exactly the same ingredients that Medina added to silver ore, in addition to mercury.[9]

Agricola's book *De Re Metallica* shows that by the year 1500 AD mining and metal-making in Europe were actually quite advanced. But most of the more sophisticated techniques and qualified people to put them to use probably did not reach the Americas until the end of the sixteenth century – with very few exceptions. Hans Tetzel, in Cuba, was one such exception (see page 16). The first water pump reached Mexico in 1575, introduced by one Cristobál Iranzo.[10] And blasting powder, which was first used in mining in Hungary in 1627, was not used in Mexico until 1703.[11] Until then, mining in the Americas was crude, involving back-breaking labour done mainly by natives who were used as slaves and worked for little or no pay. The abundance of such cheap labour was one of the cornerstones of the Spanish colonial system.

It should not surprise us that many of them died. As holes in the ground became shafts, drifts and adits, and mining continued to be carried out by inexperienced adventurers, procedures and methods had to be learned by trial and error. Real lessons were probably only learned from accidents, cave-ins, floodings and similar mishaps. The Spanish legal system contributed often to problems. One such case was the surface area a miner could claim – so small a plot that he was forced to dig down vertically, rather than horizontally or on an incline. And steep, vertical shafts increased the danger of sub-surface problems. This explains why so much of Potosí and of the Santa Barbara mercury mine was opened up vertically.

And always it was the native slave labour that suffered. Not many of the complaints were ever put in writing, and fewer still have survived. But we have detailed descriptions of work methods, and of some accidents at the mercury

Mining in the sixteenth century: the interior of a mine (although this scene is from Germany, the same methods and tools were applied in the Americas and the mines must have looked very similar). Source: Agricola, sixteenth century.

mine in Peru, which were probably fairly representative of what life in the mines was like in those centuries, almost up to the year 1800, how badly the natives were treated, how high the death rate was, how poor the remuneration and, in general, how bad their lot.

Still, we have a few numbers. Nobody counted the natives, in 1521, when Cortés reached Mexico. Estimates run from 11 million to 25 million.[12] But it is pretty clear that by 1568 the number was down to 2 650 000, and by 1603 to 1 075 000.[13] Still, even the latter numbers are only reasonable estimates as no really orderly census took place and the country was, though fully occupied and controlled by the Spaniards, not really penetrated to the last corner of the mountains and the jungles. What matters in terms of mining is that the relative stagnation of production during the seventeenth century can probably be attributed to the decline of the native – slave – population which then still formed the basic labour pool to work the mines.[14]

The search for gold and silver as well as copper and other metals was the force that drove the Spanish conquistadors to explore the remotest reaches of Mexico, and drove them soon also well into what are today the United States, Guatemala and on into Central America. Greed moved them and led to rapid settlement. Mining and the production of metals became the prime mover of Mexico's economic development.

It has already been said that attempts to put a value on the treasure, or successive treasures, that Cortés and his conquistadors took out of Mexico are meaningless. The same applies to estimates of the value of the precious metals taken out of Mexico throughout the three centuries of Spanish colonial rule. But it should be mentioned, as a curiosity, that Alexander von Humboldt, the German traveller of the nineteenth century, estimated that the total Mexican production of gold and silver bullion during the colonial period came to 2 billion pesos,[15] or about 54 000 tonnes of silver.

The Mexican mines produced much precious metal between the years 1519, when Cortés conquered Tenochtitlán, and 1821, when Mexico became independent. Technological progress in those centuries was slow, and slight. Yet the gradual technological progress did compensate for the gradual decline of grades. Most of the richest mines eventually petered out. Even as the gradual use of machines and blasting powder allowed miners to go deeper and deeper, the mines eventually became too deep to be worked further and had to be abandoned. Not even a new mining code, the *ordenanzas* of 1783, could stave off the decline, although there was a flurry of exploration and expansion between 1790 and 1810.

Already then the Spaniards knew of the existence of iron, copper, zinc, tin and lead deposits, or outcroppings, or mines. But most of these were not exploited as the metals were of no commercial value then – as and where they were. Transport was, after all, primitive and very costly.

A little copper was produced, as in Hispaniola and Cuba. But it was

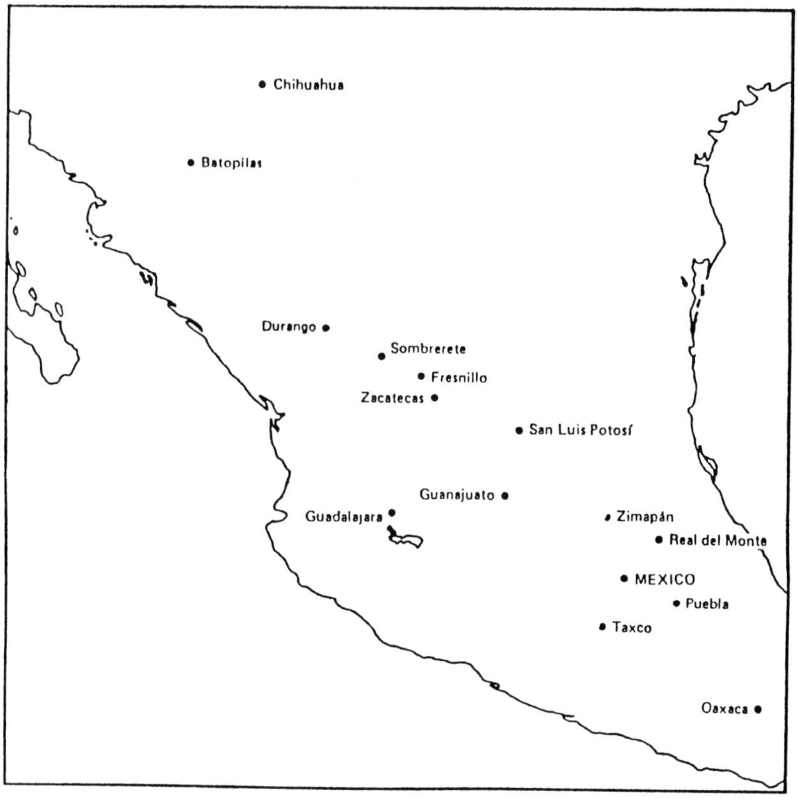

Colonial mining centres in Mexico
mentioned in the text.

generally not enough even to make the few copper pots or kettles which the sugar industry required.[16]

In 1552, the Cerro de Mercado iron ore mines outside Durango were discovered. But what a disappointment they were! The high grade ore looked like silver and the Spaniards at first thought indeed that it was a mountain of silver, but unfortunately it was only iron.

Acapulco and the Manila galleons

Substantial quantities of Mexican silver and, in time, also Peruvian silver from Potosí, went to Acapulco and from there to the Philippines. For about 250 years, between 1564 and 1815, the Manila galleons maintained a regular traffic taking American silver to the Far East to pay for silks, brocades, damasks, linen and satins as well as for spices, cotton, burlap and hemp

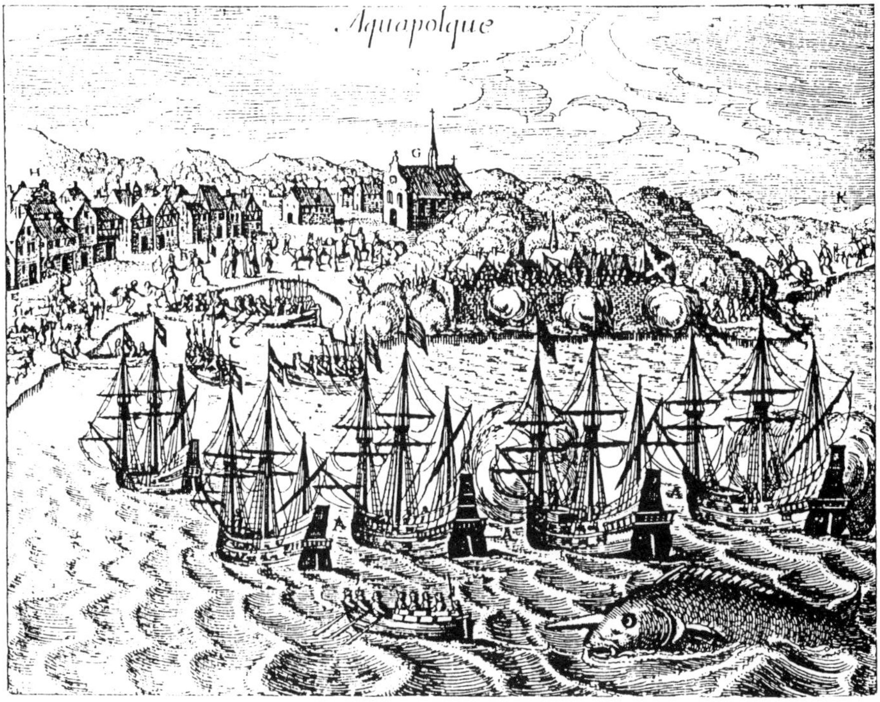

A view of Acapulco in around 1614. The Spanish galleons sailed from here for Manila. Source: Johan Ludwig Gottfriedt, *Newe Welt und Americanische Historien*, first published in 1631 by Matthaeus Merian dA and republished in 1980 as Stuttgarter Faksimile Edition im Fackelverlag, Stuttgart, Germany.

goods, and also filigree jewellery, precious stones, ivory carvings, copper and iron workings, chests and porcelain.

Some of these imports stayed in Mexico, others went down to Peru; but most went the overland route through Mexico to Veracruz and then on to Europe. In terms of mining, there was even Chinese mercury that went through Manila to Mexico.[17]

Republican times, 1810–75

The unrest during the war for independence, 1810–21, brought mining to an almost complete standstill. But it was not just civil commotion. We should not forget that many mines which had been profitable in colonial times under the then existing economic system – which was little more than slave labour – could not continue to operate with paid labour.

It was only after 1824 that a new surge in mining activity came about. Legislation which until then had barred foreigners from owning mines was changed and foreign capital was made welcome. British money was the first to come, soon followed by German and American. But political instability continued and, by 1840, production was once again regressive. The war with the United States (1846–8), the civil 'War of the Reform' (1857–61) and the French intervention (1862–7) kept mining at a minimum during these years.

The age of Porfirio Diaz, 1876–1910

It was not until the age of Porfirio Diaz (1876–1910) that peace reigned, that Mexico developed and that mining progressed substantially. Major foreign investments came into the country, helped by *laisser-faire* policies; industrialization proceeded; major railway connections were built and the country moved forward. Foreign capital was invited, favoured by the new mining code that was passed in 1884. Another law, in 1886, reduced mining taxes, further encouraging new investments. By 1895, some 40 American mining companies operated in Mexico, many in the northern states, shipping the ore to smelters north of the Rio Grande river. In the 1880s, work began at the lead/silver deposits in the Sierra Mojada, in Coahuila, at the erratic silver deposits of Batopilas, in Chihuahua, and at the El Boleo copper deposit.

The international silver price crisis of the 1890s slowed development of the Mexican mining industry once again. But a temporary recovery followed in 1906–8.

The completion of a substantial railroad network by the 1880s and its integration into the US railroad system led to an effective end of the export of lead, silver and copper ores to Europe, and thus began a new movement to smelters in Colorado, Kansas, Missouri and Oklahoma. And it also led to the building of smelters in Mexico. Railroads made the exploitation of the Coahuila coal fields possible, and the availability of coal, plus the means of transporting it, allowed the growth of a smelter industry.

And yet, the years of Porfirio Diaz were also the time when the seeds of the subsequent revolution and a resurgent nationalism were sown. Foreign investment jumped spectacularly between 1880 and 1910. British capital went from £9 million to £90 million, while US investments increased from $30 million to $1 billion.[18]

The El Boleo copper mine

Among the many interesting and colourful stories of individual mines and miners throughout the history of Mexico, one, that of El Boleo, warrants retelling.

Apache prisoners cobbing ore in northern Mexico. Source: Louis Simonin, *La Vie Souterrane* (1867).

Baja – or Lower – California is a Mexican state that is separated from the mainland by the Gulf of California. It is rugged, difficult country. There, south of the town of Santa Rosalia, a rancher one day found colourful nodules that turned out to be copper ore. The find eventually came to be identified as three bodies of rather complex ores. The grade of the first outcroppings was ranging 20–25% copper. Work began in 1872. The ore was handpicked and lightered on to ships that sailed for Europe.

In 1885, the Mexican government granted a concession to exploit the find to a French group controlled by the Rothschilds which set up the Compagnie du Boleo. They opened the mine, and built a smelter, a port, a power plant and even a 45 kilometre railroad. By 1894, El Boleo produced about 10 000 tonnes of copper matte and black copper per year. In 1924, the smelter was rebuilt and the product was thereafter known as blister copper.

The operation was highly successful and profitable for the shareholders in the early years. But by the 1930s, costs had risen, prices and grades had declined, and El Boleo's useful life seemed over. But the demand for copper during World War II and eventually tax concessions and then subsidies kept the operation going until quite recently.

Significant in the story is how a lucky find in a far-away and inhospitable

A primitive Mexican ore crusher. Source: Louis Simonin, *La Vie Souterrane* (1867).

place created, nevertheless, an industry, attracted people and, in turn, created a community. El Boleo was eventually kept going simply for social reasons.

The Guggenheims

Among the names that loom especially large in the history of the Mexican mining industry is Guggenheim.

Simon Meyer Guggenheim, the patriarch of the family, had, at age 55 taken his children and those of his second wife from her first marriage from Switzerland to America to make a new start – and that meant as a pedlar in Philadelphia. His son, Meyer Guggenheim, did likewise. Their early staples, and successes, were stove polish and coffee. They profited from trading during the American Civil War and, once they had enough capital, became wholesale merchants in spices. They next ventured into the production of lye. Successful speculation in railroad stocks gave them sufficient capital to import laces and embroideries from Europe.

In 1881, Meyer Guggenheim made a blind investment, sinking $5000 into a one-third interest in two paralyzed lead/silver mines in Leadville, Colorado. It was a real flyer.

He soon travelled to Colorado to take a look – and decided to buy out the partners and acquire more equipment. The mines had to be pumped dry and repaired.

But once work began, they struck it rich. And the investment – by then probably about $70 000 – brought a quick and steady return of $17 000 a month. By 1888, the mines gave about $750 000 profit for the year.

The next step was to build their own smelter, at Pueblo, Colorado. Soon they were importing ores from Mexico, for blending, and this led the Guggenheims – for Meyer Guggenheim had by now his seven sons in the business – into the Mexican mining industry.

In 1890, the US Congress passed the Sherman Silver Purchase Act which caused the silver price to rise from 90¢/oz to $1.25/oz. At the same time, the McKinley Tariff Act penalized imported ore heavily. The result was that the Guggenheims had unexpected excess profits from their silver production in the United States. They now decided to build smelters in Mexico to process ores near the mines.

They took quick decisions, leasing mines in Nueva Leon and Jalisco and building smelters at Monterrey and Aguascalientes. In the characteristic Guggenheim fashion – direct, decisive – Daniel, one of Meyer's sons, went to Mexico, negotiated with Porfirio Diaz himself and obtained the necessary leases and permits. Even the machinery for the mines and smelters could enter Mexico duty-free.

Construction proceeded rapidly. By the end of the fourth year of operations, the Monterrey smelter had repaid its investment. By 1895, the Guggenheims were a major industrial force in Mexico.

More investments followed, more mines and smelters, at San Luis Potosí and Veracruz, and especially the rich gold mine Esperanza.

In 1913, the *Engineering & Mining Journal* reviewed the silver, lead and copper smelting works of North America and found that there were three blast furnaces in Canada, 76 in the United States and 45 in Mexico. By then the interests of the Guggenheims and those of the Rockefellers had merged into a large trust, the American Smelting & Refining Company, to be known as ASARCO; the Guggenheims came to be in charge of ASARCO, and ASARCO controlled half of the non-ferrous smelting capacity of Mexico.

Another significant step in the development of the Mexican mining industry was the gradual introduction, around the turn of the century (1895–1910), of the cyanide process. Much cheaper than the old patio process, the cyanide process was introduced after the availability of cheap electricity made large scale fine milling possible. The mines could now process even lower grade ores, and a new boom in silver production followed.

Despite the slowdown between 1906 and 1908, metal production almost doubled from 1900 to 1910, mostly because of the influx of foreign capital, know-how and technicians.

The revolution, 1910–20

But much was wrong in Mexico by then. Workers, and especially mining workers, were poorly paid, foreigners dominated too much of the mining and other industries, and, most important, the Porfirio Diaz regime had already been in power 35 years and was in many ways corrupt. As nationalistic feelings grew, so did xenophobia. Yet a new mining law, passed in 1910, was, nonetheless, still fair and reasonable to foreign capital, and the boom appeared likely to continue.

Revolution, however, interfered and much turbulence reigned between 1910 and 1920. It became a decade lost. In 1910, Francisco Madero succeeded Diaz. When Madero was assassinated, in 1913, he was succeeded by General Victoriano Huerta. But there were also others, like Venestiano Carranza, the governor of Coahuila, the bandit chief Pancho Villa, General Lazaro Cárdenas, the peasant leader Emiliano Zapata, and Alvaro Obregón.

By 1914, Huerta went into exile and Carranza became president, to be assassinated in 1920. Zapata had already been assassinated in 1919. Obregón became the next president.

The ups and downs of the revolution often degenerated into civil war. Much of the mining industry suffered, but not all. El Boleo was not affected, but many of ASARCO's mines and smelters stopped operating. Production was down 75%.[19] 'Almost every operating mine was held up at one time or another with demands for horses, supplies and hard currency.'[20] Much of the railroad system collapsed, as bridges were destroyed and rails torn up. Yet there were also efforts by the revolutionaries to avoid molesting foreigners and their properties.[21]

World War I was actually a good time for mining operations as metal values rose. But while El Boleo benefited, most of the silver mines did not as the import of German cyanide was interrupted. And the United States embargoed dynamite shipments, which action alone was estimated to have caused production to have fallen by half.[22]

Coinage dropped to zero and Mexico, producer of about one-third of the world's silver, had suddenly not enough specie.[23] The country went on a paper standard and bills were literally printed by anybody who had access to a printing press, including Pancho Villa. Some bills were even mimeographed.

The losses of the mining industry were enormous – equipment was destroyed, some mines flooded, railroads torn up and much production lost. Revolution not only meant unrest and military action, robbery and assassination. It also brought new tax laws. Mines were ordered to resume work, to perform a social service and to produce value on which taxes could be levied. Villa even issued decrees to the effect that mining was a public utility that also provided work for the poor.[24]

Considerable confusion reigned in the years 1915 and 1916. Carranza, pressed by the need for cash, passed new tax laws. Some were meant to collect more money, others to force owners to put idle claims into operation, yet others to penalize large holders – clearly meant to affect foreign companies – and to help small, usually Mexican, owners. The disputes became bitter; the US State Department intervened, demanding annulment. But Carranza was unyielding.[25] It was obvious that the laws were indeed meant to reduce foreign holdings.

In 1917, Mexico got a new constitution. It was populist, re-establishing the state's supreme right to mining property and giving labour equal rights with mine owners. It was not until 1926 that a new mining law came about.

From 1917 on, the industry began to pick up again, yet by 1919 still only 12% of the mines and 21% of the smelters were operating.[26]

Boom and bust, 1920–38

New tax laws in 1918, fewer restrictions on foreigners seeking concessions and increasing domestic tranquillity caused production to rise again. Prices continued high as World War I reached its climax. The United States had finally entered the war in 1917, and its industry was now running in high gear. More government income brought efforts to retire most of the worthless paper money.

A short depression in 1921–2, was followed by good years when mining again expanded. But the unrest and excesses of the previous decade had reduced foreign interest in investing heavily in the Mexican mining industry again. The rampant nationalism of the revolutionary years had also had negative effects. Still the industry grew, in line with the general boom in metals throughout the 1920s. While few new mines were discovered, some old ones thrived, among them Cia. Minera Peñoles which in 1920 merged with Cia. de Minerales y Metales to become one of the largest groups, second only to Guggenheim/ASARCO.

And then there was selective flotation. Until the early 1920s, complex base metal ores could not be treated. But with the new flotation process, different finely ground metal particles would cling to different oils and chemicals when mixed with air and water, and the metals could then be recovered separately. The process finally allowed for profitable extraction of zinc. Base metals production now really took off.

Production statistics for the period 1890–1990 tell the story of the fluctuating fortunes of metal mining in Mexico in the period.

Zinc

Only minor quantities of zinc were produced until 1905. They were probably just calamines. In 1906, production suddenly rose to 23 000 tonnes, but it was only World War I demand which caused production to jump to 45 000 tonnes in 1917. It then fell again to 1200 tonnes in 1921. However, with selective flotation, production of zinc rose after 1925 to a high of 175 000 tonnes in 1929, a peak that was only to be surpassed in 1942, again with war demand, to rise thereafter to an all-time high of 295 000 tonnes in 1955. Since then, production has fluctuated but has always stayed safely above 200 000 tonnes per year.[27]

Lead

The Mexican statistics show that as much as 3 million tonnes of lead were produced between 1521 and 1890,[28] by which time production was at 30 000 tonnes a year. It then fluctuated widely, between a low of 5000 tonnes in 1914 and a high of 120 000 tonnes in 1910, to then rise to 240 000 tonnes in 1927, and to an all-time high of 248 000 tonnes in 1929. Never again did lead production go that high. By 1980, production was down to 147 000 tonnes, to recover again to 184 000 tonnes in 1983.

The big jump in the 1920s, when production doubled to the 200 000 to 300 000 tonnes range, was, beyond doubt, due to the introduction of selective flotation.[29] It was a major leap in the development of Mexican – and worldwide – mining. Mexico now became the second largest lead producer in the world, and one of the largest of zinc. Yet no new mines were discovered.[30]

Copper

Yearly production of copper was around 11 000 tonnes in the 1890s. But it quickly doubled by the year 1900, and again by 1903, and reached 65 000 tonnes in 1905, a level it was not to see again until 1925. The fluctuations thereafter were within quite narrow limits and production was only to reach 80 000 tonnes in 1973. Output rose steadily in the 1980s, to the 250 000 tonnes/year level at the end of the decade.

Silver

Production of silver fluctuated from 1100 tonnes in 1891, to a low of 712 tonnes in 1915 during World War I (undoubtedly attributable to the shortage of cyanide), and to a high of 3400 tonnes in 1929. It then gradually declined over the subsequent decades to a low of 1150 tonnes in 1965 before rising back to 2300 tonnes in 1989. (Mexican statistics reveal that some 84 000 tonnes of silver were produced between 1521, when the Spaniards arrived, and 1890.)

Gold

Gold production showed a similar pattern. Production from 1891 on reached a high of 41 tonnes in 1910, from a low of 1.5 tonnes in 1891, to then decline steadily (with drastic fluctuations from 1914 to 1916) to a new low of 4.1 tonnes in 1913, before rising again to reach 9.6 tonnes in 1990.

Yet the most astonishing statistics about the Mexican mining industry in the period 1910–40 are that foreign interest continued to control 97% of it in 1910 and 98% in the mid-1920s.[31]

Difficulties were not far away. President Plutarco Calles (1924–28) was xenophobic, compared with Alvaro Obregón who had turned out to be quite pro-mining.[32] Calles put out a new mining law in 1926, finally applying to mining the basic provisions of the 1917 constitution.

Reading the 1926 mining law now reveals what appears to be a modern, normal and reasonable document. That Mexicans were given preference in equal circumstances, that foreigners could participate in mining only under a number of restrictions, that employees and labour had to be at least 90% Mexican, that a claim holder had to keep up a certain level of work – and other, similar provisions – all sounds very reasonable today (when such provisions apply in many countries). But they were probably downright revolutionary and objectionable then, in 1926, and in a country where foreigners still controlled at least 95% of all mining operations.[33]

Trouble now intervened. Silver prices tumbled from 66¢/oz in 1926 to 53¢/oz in 1926. New tax laws came out once more, in 1926, and were very harsh. They had to be modified only a few months later. More tax changes were then made in 1927 and 1929, in both cases to lighten the burden on the industry once more.

Lower metal prices in the late 1920s gradually made more and more mines marginal and forced companies out of business. The drop in production after 1930 was precipitous. It was, of course, also the time of the depression.

By 1932, lead and copper production had fallen by half, and zinc by two-thirds, from their high levels in the late 1920s. Gold production, curiously enough, only fell by about 15%.

By about 1934, the trend was up again, led by silver. A real silver boom developed when the US Treasury, beginning early in 1934, commenced buying silver massively. For a while, the price rose from a low of 45¢ to 81¢, only to then settle again at 44¢/oz.

Nationalism and World War II, 1938–50

A new president, Lázaro Cárdenas, came to office in 1935. His orientation was leftist, nationalistic and xenophobic. In 1938, he nationalized the foreign oil companies. The foreign mining interests thought they would be next; but it did not happen. The metal mining companies were left alone. But the damage had been done; distrust kept foreign capital away and the production highs of 1929 were not reached again[34] (except for zinc in the late 1950s and copper in the 1980s) despite the demand caused by World War II.

We should take another look at the overall production statistics, as they tell the story quite clearly.

World War II brought a new boom. At first, trade patterns were interrupted, especially in zinc and lead which had gone principally to the European market – 80% of lead and 70% of zinc. In fact, Germany had been the largest market, taking, in 1939, 19% of Mexico's lead and 26% of its zinc.[35] Stocks at first accumulated, prices dropped and some of the mines, including those of ASARCO, suffered. Peñoles even had to shorten the working week.[36] But gold, silver and copper continued to flow well into the US market.

Mexico's new president, Manuel Ávila Camacho (1940–6) proved to be less nationalistic than Cárdenas. But he still meddled with ore marketing, regulated mill and smelter charges and saw to generally stricter enforcement of existing laws.

In July 1941, Mexico and the United States signed an agreement, valid for 18 months, that all metals and ores not sold within the western hemisphere would be bought by the United States. By the middle of 1942, further negotiations also settled all pending oil claims. Pearl Harbour, of course, also led to a strengthening of ties between the two countries, including a new agreement providing for massive purchases of Mexican metals by the United States at rather high prices. As a consequence, a new boom in mining developed.[37]

The Mexican government promptly raised taxes and everybody benefited. But when the war ended, US purchases dropped, prices collapsed and so did production in Mexico. In November 1945, the United States cancelled all pending contracts; mines began to close[38] and unemployment rose.

A new boom, though smaller, came at the time of the Korean war. Mexico again benefited, but in general the industry entered into a long cycle of gradual decline.

A declining role for mining 1950–90

The overall role of mining in the Mexican economy has declined gradually throughout the last four decades. While there have also been a few occasional ups, the overall impression is one of decline. Mexico has, of course, been

Table 3.1. Mining production figures, 1953–91 (tonnes)

	Gold	Silver	Lead	Zinc	Copper
1953	15.0	1459	221 000	252 000	60 000
1958	11.0	1460	198 000	248 000	68 000
1963	7.0	1297	184 000	241 000	55 000
1968	6.0	1245	174 000	240 000	61 000
1973	4.0	1206	167 000	271 000	81 000
1978	6.0	1579	171 000	245 000	87 000
1983	7.4	2012	184 000	266 000	196 000
1988	10.7	2358	171 000	262 000	268 000
1989	11.5	2306	163 000	284 000	249 000
1990	9.6	2346	180 000	299 000	291 000
1991	9.8	2290	168 000	317 000	267 000

changing – manufacturing has grown and the country has become highly industrialized. While mining was somewhere around 10% of GNP before World War II, it declined to 2.5% by 1955 and is probably today below 2%. But there seem to be other reasons for the decline, of a more fundamental nature.

Labour has become too powerful and thus too inflexible, over-protected by anti-capital laws. Not too much exploration has taken place and thus few new mines have been discovered; still fewer have been opened up. Mexico was warned, in 1953, by a report of the International Bank for Reconstruction and Development that erroneous government policies were seriously affecting the mining industry. Still, from that date on, mining production declined as shown in Table 3.1.[39] The reader may wonder why these statistics seem to prove a decline as in three of the five metals the totals are indeed up. But we must look at these figures in terms of the overall growth of the economy, and especially of the population, which in these 40 years just about tripled.

Mexico must probably continue its tortuous road towards a fully modern, open, investment-friendly economy; as long as it does not, the decline of its important mining industry will continue.

Could progressive 'mexicanization' have played a role in the decline? We already saw that in 1910 Mexicans only owned 3% of the mining industry of their country, and even in 1925 only 2%, obviously an absurdly lopsided relationship. But by 1961, Mexican ownership was already at 10%. And then came the new mining code of 1961.

This new basic law established that Mexicans would henceforth hold majority ownership of all mines and of the management of mining companies. Though the law gave 25 years for foreign companies to become mexicanized, the process went much faster, and by 1972 Mexicans controlled practically 100% of the industry.[40]

Table 3.2. Silver production, 1521–1990

Period	Tonnes
1521–1900	99 343
1901–76	143 810
1977–89	25 500
1990 alone	2 346

The Mexicans claim that after the 1961 law came into effect[41] and the mexicanization began, major investments were made in exploration and major new deposits were discovered. Should not enough time have elapsed by now for us to see the results?

It should, however, be said that by 1992, under President Carlos Los Salinas Gortari, the country had opened up once more, creating more favourable laws (for foreign investors) and a more hospitable investment climate. Maybe the wheel is turning the other way once more.

One additional comment on silver production throughout the centuries will bring into focus what was said at the beginning of this chapter about the fabulous treasure the original conquistadors took out of Mexico. The treasure was indeed magnificent. But in terms of overall quantity mined since, was it significant? Table 3.2 shows the production figures for silver, over the centuries. And the value of the other metals (other than gold and silver) has by now far surpassed the value of gold and silver.[42]

It seems peculiarly fitting to conclude the chapter on mining in Mexico with a curious twist, or turn, of fortune: by 1992, Mexico was importing mercury, to the tune of about 5000 flasks per year. What has happened? The country which was once foremost among the mercury-producing nations of the world is now an importer.

The explanation is quite simple. The business had continued in the hands of small miners, called *gambusinos*; they had generally operated without much, or any, government control, technical or fiscal. When the government began to clamp down, in recent years, the *gambusinos* went out of business. Imports are now satisfying the demand which still exists.[43]

Notes

1. López de Gómara, *Historia*, volume 2, page 56.
2. At 230 grams per marco (mark). See Purser, *Metal Mining in Peru*, page 317, also Hemming, *The Conquest of the Incas*, page 518. They say 1 marco contained 50 castellanos or gold pesos, which, at 4.55 grams per peso would bring the weight of a marco to 227 grams.

3. Arciniegas, *Biografica del Caribe*, page 141.
4. Ibid, page 142.
5. Bernstein, page 9. Just as Riverend's *Historia Economica de Cuba* is invaluable for any review of mining in Cuba, the history of the mining industry in Mexico cannot be understood without Marvin D. Bernstein's *The Mexican Mining Industry, 1890–1950*. It is a thoroughly researched book which allows us to understand the full scope of society, industry, social movements and politics. By also covering, in the introductory chapters, the history of mining since the Spanish conquest, the book is really a complete review of the history of Mexico. I acknowledge that without Mr Bernstein's book, the interaction between the general history of Mexico and the role mining played in it could not be understood.
6. Ibid.
7. Ibid, page 10. I am essentially following the outline given by Bernstein.
8. Bernstein makes similar comments about the pre-Columbian natives of Mexico. He also describes, on page 7, how rich some of the original veins must have been. If indeed some veins ran as high as 75% silver, not much mining technique was required.
9. Agricola, *De Re Metallica*. For gold from ore, see pages 347–9; for amalgamation, see page 209.
10. Semo, *Historia del Capitalismo en Mexico*, page 46.
11. Ibid, page 39.
12. De la Peña, *La Formacion del Capitalismo en Mexico*, page 53.
13. Semo, pages 29–30.
14. De la Peña, page 53.
15. Bernstein, page 12. I have already discussed smuggling and suggested that as much as 50%, if not more, was often smuggled or stolen. The real amount could thus be double Humboldt's estimate. Still, 2 billion pesos (of silver, at 27.468 grams per silver peso) would give as much as 54 000 tonnes of metal.
16. Konetzke, *Fischer Weltgeschichte*, page 300.
17. Lyon, Eugene, in the *National Geographic Magazine* of September 1990.
18. Cheetham, *A Short History of Mexico*, page 200. But Bernstein puts US investment in 1914 at only $300 million. 'Estimated value' versus 'registered investment'?
19. Bernstein, page 100.
20. Ibid.
21. Ibid, page 97.
22. Ibid, page 102.
23. Ibid.
24. Ibid, page 107.
25. Ibid, pages 110–11.
26. Ibid, page 118.
27. The statistics come from the Secretaria de Energia, Minas y Industria, Paraestatal, Mexico, DF.
28. Ibid.
29. The statistics really speak for themselves.
30. Bernstein, page 139.

31. Ibid, page 143.
32. Ibid, page 149.
33. 'Foreign' really meant 90% US, British and French.
34. Bernstein, page 185.
35. Ibid, page 223.
36. Ibid, page 224.
37. Ibid, pages 224–7.
38. Ibid, page 223.
39. Again, these figures are from the basic statistical sheets of the Secretaria de Energia, Minas y Industria, Paraestatal, Mexico, DF.
40. León-Portilla, *La Mineria en Mexico*, page 174.
41. Ibid, page 175.
42. Ibid, page 169.
43. Verbal communication from Victor Vazquez, November 1992.

CHAPTER 4

Central America

Guatemala, El Salvador, Honduras, Nicaragua and Costa Rica

Mexico had not yet been completely conquered when the Spaniards began to push south. The conquest of *Guatemala*, in 1523, is generally credited to Pedro de Alvarado. Little gold was found and they pushed on. Since then, lead, zinc, copper and silver, in addition to some gold, have been found, even some antimony. The production of the latter reached, in 1987, the respectable quantity of 1880 tonnes.

It was again Pedro de Alvarado who, in 1524, conquered *El Salvador*, once more in search of gold. But he found little. However, in time gold and silver were found and also, more recently, iron ore, mercury and lead.

In *Honduras*, gold, silver, lead and zinc have been mined off and on. Zinc production reached a peak of 44 000 tonnes in 1985; lead production declined in the late 1980s. Copper, tin, iron and antimony also exist, with small and sporadic productions.

Nicaragua also has gold, copper and silver as well as deposits of lead, iron and zinc.

It was Columbus himself who named *Costa Rica* the 'rich coast' because some natives he met when reaching the coast wore golden adornments and he thought untold wealth lay inland. The Spaniards never found much gold, but eventually some was found and still today gold is listed as the principal mineral export of the country.

Natives delivering jewellery and adornments as tribute in Panama in around 1514. Source: Johan Ludwig Gottfriedt, *Newe Welt und Americanische Historien*, first published in 1631 by Matthaeus Merian dA and republished in 1980 as Stuttgarter Faksimile Edition im Fackelverlag, Stuttgart, Germany.

Panama

Panama – in colonial times called Darien – is slightly different from the other Central American countries. Rodrigo de Galván Bastidas discovered it in 1501. In 1510, Diego de Nicuesa founded Nombre de Dios on the Caribbean side. And then Vasco Nuñez de Balboa saw from its shores the Pacific Ocean in 1513, and from Panama city – the original settlement, a short distance from today's city – Pascual de Andagoya in 1516 and Francisco Pizarro in 1524, set sail, south down the west coast of South America, eventually leading to the discovery of Peru.

The Panama of colonial times apparently had considerable quantities of placer gold in its jungle, especially the area near the Colombian border. No wonder, as subsequently major quantities of gold were found in the adjoining Colombian lowlands.

Fray Bartolomew de las Casas, the protector of the natives and early chronicler of the conquest and settlement of the Caribbean, tells us of a 'great chief', called Paris, who lived some 10 or 15 leagues away from Panama city and who was very rich, especially owning a lot of gold. One day Paris was visited by a group of Spaniards. He supposedly treated them like brothers and gave them golden presents. But the Spaniards were treacherous and fell upon Paris and his people and burned the native village. Paris managed to escape and, three or four days later, he in turn fell upon the Spaniards, killed 50 of them and recovered his gold, worth some 140 000 pesos or, in our terms, some 630 kilograms. The Spaniards soon returned, however, with a stronger force and Paris and all his people were killed.[1]

Many native lives must have been lost in Panama when the Spaniards worked the rivers in what is still today inhospitable jungle, so forbidding that it forms the only natural barrier – or interruption – to the continuation of the Panamanian highway which spans both continents.

It was from Nombre de Dios and Portobelo – and from Hispaniola and Cuba – that the first exploration of Colombia's Caribbean coast began, just as it was from Panama, on the other side of the isthmus, that exploration began down the Pacific coast.

We should maybe once again pause to visualize what life must have been like in those days. The original town of Nombre de Dios has disappeared, as has Portobelo. They must have been then, between 1510 and 1710, fever infested trading posts as well as luxurious administrative centres – especially after all the wealth of Peru and Bolivia started to come through. And the same must have applied to Panama city, on the Pacific side. One cannot but marvel when seeing today the ruins of Old Panama, the original settlement, long since abandoned.

A Scottish foray into Darien

In about 1680, the itch for adventure and to perhaps participate in the easy pickings of gold in the Caribbean also caught the imaginations of some people in Scotland. A company was formed and, in 1698, 1200 sturdy Scots set forth, in four vessels. Many of those on board did not even know where they were going.

The Scots were not modest in their aims. Their objective was Darien, the isthmus of Panama, to get control of the flow of wealth from the Pacific, to settle the bridge between the two continents as England had settled, or was settling, New England. This now was to be New Scotland, and their town was to be New Edinburgh. They settled somewhat east of Nombre de Dios.

But the English disliked the idea of yet another party entering the power-plays of the Caribbean. And the Spaniards obviously objected to the intrusion in what they considered their territory, the 'Spanish Main'. They put together an army to battle the Scots.

But the Scots were already not doing well at all. Heat, humidity, fever, hunger, insects – all came together to defeat them. A vessel they sent back to Barbados foundered on some rocks and the crew ended up in Spanish jails. Others, fed up, left for Jamaica and New York. They were the lucky ones. The remainder feebly battled the Spaniards and were then allowed to withdraw honourably, in several vessels. One of these hit some rocks off Cuba, and two others were destroyed by a hurricane when at anchor in the port of Charleston, South Carolina. Very few of the settlers returned to Scotland. The dream was soon forgotten.[2]

Travel across the isthmus was difficult at first and took weeks. However, with time and regular paths, and as mule trains replaced human porters (native slaves) the time shortened considerably. The insect infested swamps became less threatening. But a joyful crossing it can never have been. Only once a railroad was built did travel become more amenable. Illness remained a major problem and, still in 1880, when the French attempted to build a canal, thousands died of yellow fever.

Gold and silver have always been mined in Panama. There is also manganese, quite a bit of copper, and small quantities of bauxite.

We should now turn towards South America . . .

Notes

1. De las Casas, *O Paraiso Destruido*, pages 46–7.
2. Arciniegas, *Biografia del Caribe*, pages 287–303.

CHAPTER 5

South of the Spanish Main

Venezuela and Colombia are today separate and independent countries, as are Equador, Guyana and Surinam. It was not always that way. The modern history of this area, the northern, north-western and north-eastern corners of South America, began with their discovery by the Spaniards in 1498, when Columbus, during his third voyage, landed on what is today the island of Trinidad.

What there was before

What we know of what existed and happened before the arrival of the Spaniards offers nothing comparable to Mexico, Guatemala and Peru. A number of native tribes lived in the eastern part of the area, mainly occupied with hunting, fishing, gathering and some planting, and without much of a higher civilization. The spread of these tribes apparently also covered what are today the Guianas and reached all the way down to the Amazon lowlands.

Further to the west, in today's Colombia, the tribes were not very different except maybe for the Chibcha in the Andean highlands. But we may simply not know enough as yet.

The history of the pre-Columbian people in both North and South America remains confused and uncertain. New discoveries are made all the time and the general picture changes every few years.

For instance, in 1975 gold looters discovered the remnants of a 1500 year old city in the mountains on Colombia's Caribbean coast; it became known as the 'Lost City of the Tairona'. In the nearby lowlands 'the most extensive pre-

Columbian irrigation system known to exist in the Americas' is being checked out. And near the town of San Jacinto, also close to the Caribbean coast, 'some of the oldest pottery in the New World', dated to 3990 BC, was discovered.[1] And in south-western Colombia, near a town called San Agustin, some 500 stone statues are scatterd over a 15 square mile archaeology park. They seem to date from about the time of Christ.[2]

If man did indeed come to the Americas via the Bering Strait, from Siberia, and walked down the west coast of the Americas, Panama and north-western Colombia must have been crucial stepping stones on the way. Bogotá's Gold Museum and other museums in the world hold innumerable intricately worked and unbelievably beautiful pieces of Colombian gold. And more pieces come to light each year. Yet probably no more than 30% of all finds come on to the (illegal) art market; most is probably melted down and enters the partly legal but mostly illegal (smuggled) bullion trade.

In time, major centres of pre-Columbian activity were identified, foremost among them the Tairona region near Cartagena, on the Caribbean coast, the Chocó just south of the Panamanian border, the Muisca around modern-day Bogotá, and the Quimbaya west of Bogotá. The Calima civilization was named after the valley of the same name, located just south of Quimbaya. Calima has become famous for its rich graves – and all these civilizations for their wonderful golden artefacts.

Gold is also why the Spaniards came.

The history of metal mining in Venezuela and Colombia is first of all the search for El Dorado, the Golden One, or the Golden Man. It is also the story of the pearls from the coastal islands, the gold rush of the Chocó, the mines and placers of the Caucá valley, of the Guiana jungles, of the fabulous emeralds, and of the iron ore and bauxite mines that are being exploited today.

There was, and is, gold all over these countries. In 1988, Colombia produced 33.4 tonnes of gold and was thus the second largest producer in South America and the fourth largest in the western hemisphere. Venezuela produced 16 tonnes of gold which made it the fourth largest producer in South America, behind Brazil, Colombia and Chile. Most of Venezuela's gold comes from the Guiana province. The country also has copper, much iron ore, some lead, zinc and antimony, as well as large quantities of bauxite.

Again it was Columbus who, in 1498, during his third trip, touched the shores of modern Venezuela for the first time. Alonzo de Ojeda followed in 1499 and he called it Nueva Andalusia. When he entered Lake Maracaibo, he called the region Venezuela, as it reminded him of Venice. One of Ojeda's companions, a Florentine merchant by the name of Amerigo Vespucci, was a few years later to claim to have been the leader of the trip, as well as of a subsequent one to Brazil which, in 1509, led a geographer to call the entire continent America.[3]

We must visualize that 'Tierra Firme' was but a short distance away, sailing straight south from Hispaniola which was, at least in the early years (1493–1530), the centre of all Spanish activity in the Caribbean. It was first called 'Tierra Firme', then Castilla del Oro, but soon, mostly with the coming of the Welsers, (see below), a distinction began to be made between the Audiencia de Santa Fé (de Bogotá), part of the Viceroyalty of Peru, and the Gobernación de Venezuela, part of the Audiencia de Santo Domingo which, in turn, was part of the Viceroyalty of New Spain – Mexico.

After a few confused attempts at settlement and several expeditions inland, the Spanish efforts were temporarily overshadowed by those of the Germans. But first came the pearls . . .

Pearls from the islands

Ojeda was followed by Pedro Alonso Niño, Luis Guerra and Christobal Guerra who visited the islands of Margarita and Cubagua in 1500 with one caravel and 31 men to search for pearls. Niño had studied Columbus' reports and had picked out the reference to pearls. The three and their group found pearls. On their way back to Spain they put into Santo Domingo, for provisions and stores, and their stories led immediately to a stampede. Hordes of adventurers now fell on the islands and for a while it was a lucrative yet deadly business. They, the Spaniards, would not dive and search for the oyster beds. This had to be done by natives – and we again read tales of working the native slaves literally to death to get the pearls.[4] After the year 1600, the conch beds became exhausted.

The Welsers – German bankers intervene

It was in January of 1528 in Burgos, Spain, that representatives of the German banking and trading house of Welser signed a contract with the Spanish crown. It gave the Welsers the governorship of as well as the rights to explore that part of South America which later was to become the General Captaincy of Venezuela, then the Audiencia de Santa Fé and eventually the Viceroyalty of New Granada – and still later the Republic of Venezuela.

The Welsers were the second most powerful – next to the Fuggers – trading and banking house of Augsburg, Germany. Their main activities were spices and general trading.

In 1520, after Maximilian of Habsburg died, it was Fugger and Welser money which ensured that Maximilian's grandson, King Charles I of Spain since 1516, also became Charles V, Emperor of the Holy Roman Empire of the German Nation. And as Charles owed favours to his bankers and financial backers, the governorship of 'Little Venice' (named thus because at first there seemed to be only islands) – Venezuela – was in payment of one of those IOUs.

The Welsers were neither miners nor metal traders, the way the Fuggers were. But they thought they had a good nose for business, and they were anxious to steal a march on the Fuggers. The contract 'to explore and exploit' Venezuela was thus, they thought a royal – no, an imperial – licence to participate in the search for and enjoyment of the wealth of the New World. Columbus had shown the way, but it had been the quantity and quality of the gold from Mexico, pouring into Spain and Europe from about 1520, which had really excited the public.

In 1528, the Welsers sailed for Hispaniola to establish an office and to prepare their conquest of the mainland from there. A year later they landed in Venezuela and founded a small town at the southern base of the Paranaguá peninsula. They called it Coro. Their leader and governor of the colony was one Ambrosius Dalfinger, but as the Spaniards had great difficulty pronouncing German names, he became Micer Alfinger. He did not linger in Coro; soon he was off to explore – and to search for gold.

His direction was west, to where rumour pointed to another sea, and more gold. And both proved correct – more or less. Alfinger found Lake Maracaibo, indeed a large body of water, and gold, jewellery, and other adornments, beautifully worked. The explorers enquired where the gold pieces came from – as the excellent workmanship was obviously not of the local tribes – and were told 'from a people high up in the mountains, further inland', where there were also large quantities of beautiful green stones. The Spaniards, or Germans,[5] were also told that these very same highland natives also had odd animals, similar to sheep, which turned out to be llamas.

The search for El Dorado

These disjointed pieces of news, enlarged and embellished by the retelling, and proven by more and more pieces of gold, must be looked at as the origin of the legend of El Dorado, the Golden One. One year earlier, in 1528, Francisco Pizarro had established the first contact with the Inca empire, when he reached Tumbes in northern Peru. This first visit was only tentative, but Pizarro and his people also heard of much gold, of a 'sun empire', and their stories, told in Panama and elsewhere in the Caribbean, and again told and retold in Spain during Pizarro's visit in 1528–30, excited everybody.

The news which thus came from the interior of Venezuela (and Colombia) with Alfinger upon his return to Coro in 1530 coincided with similar stories from elsewhere.

El Dorado referred originally to 'the Golden One', or the 'Golden Man'; only later did it become the concept of a country, or an empire, or an area, Eldorado. The reference to a golden man was not pure fabrication; there was a factual basis to it. Historians have been able to piece together that the Chibcha Indians, a small and minor tribe in the context of pre-Columbian

America, were living in the general area of modern-day Bogotá – the inter-Andean highland of today's Colombia. The Chibchas were neither warriors nor artisans; theirs was not a large country, nor were their villages cities. All they had was salt – the salt mountain of Zipaquirá. Large salt mines were what made them unique and gave them some power and some wealth. They also had emeralds, as the Muzo emerald mine was in their area and was being worked by them. The Chibcha traded the emeralds and the salt for other goods, mainly gold.

However, the stories that reached Coro in 1530 spread on. And the Welsers now sent expedition after expedition into the interior.

When Alfinger stayed away almost a year, Augsburg appointed a new governor, Johannes Seissenhofer, who, however, died soon after reaching Coro. His second in command, Nikolaus Federmann, now became governor as Alfinger, when he finally returned to Coro, was too ill to administer anything; he quickly left for Santo Domingo to convalesce. Federmann, a flaming redhead, soon succumbed to the lure of gold and also set out for the interior.

In the reports that relate the arrival of Seissenhofer and Federmann we find an occasional reference to 'miners', which note that they, the German miners, were useless in a country where there were no mines. And though they found iron and manganese deposits, the tools and other means to exploit them were totally lacking.

Federmann's expedition lasted six months. Not much was achieved, except that they again obtained many golden objects and new stories. When Federmann returned to Coro, in March of 1531, he found Alfinger already back from Santo Domingo, well recovered and again fully in charge. Alfinger resented Federmann's trip into the interior and now banished him from Venezuela for four years.

By September 1531, Alfinger was off once again. Now it was a longer trip and it led him all the way to the Magdalena river, which he then followed up river for a while. The story of the trip is similar to those before, and those that would follow. Some of the natives they met were friendly, others not. Some supplied them with food, others had to be battled with. And everywhere there was gold, and more stories about more gold further up river, up country. Like the other groups, this one also suffered – from hunger, thirst, illnesses. At some point, Alfinger decided to send back the accumulated treasure: 1723 pieces of jewellery, and some 110 kilograms of gold. Twenty-nine men accompanied the treasure; only one reached Coro.

In June of 1532, another group was sent back, again with accumulated treasure and an urgent message to send supplies and reinforcements. The rest pushed further south, up river, up country, deeper into the Andes. Cold, ice and snow now became as unbearable as insects and heat had earlier.

Eventually, Alfinger decided to turn east, and then back north. The expedition was now constantly under attack by hostile natives. During one of those battles, Alfinger was wounded by a poisoned arrow; the curare took its course and a few days later they buried him. A few months later, 35 of the original 175 men made it back to Coro. It was November 1533. They had been away 29 months.

The results so far were poor – some gold, much hardship, many deaths. But the Welsers did not give up. Only a few months later, in January of 1534, the galleon Santa Maria del Campo reached Spain with the story of Cajamarca, and how Pizarro had vanquished the Inca army, how he had imprisoned Atahualpa, and what fabulous ransom the Inca had paid. And part of the ransom was there, in Seville, for all to see.

The Welsers now sent yet another German with an unpronounceable name to Venezuela to take over as governor: Georg Hohermuth, from Speyer. They called him Jorge Spira. He reached Coro in October of 1535. Federmann also reappeared. His four years of exile had ended. He was the first to start out on the already mandatory expedition into the interior, but it was only a short expedition – straight west toward Santa Marta. But that was another province, already occupied by Spaniards and not part of the Welser territory. The local governor, Fernandez de Lugo, forbade Federmann to continue and, as Lugo had more troops, Federmann gave up and returned to Coro.

Meanwhile, Hohermuth had left on what was to be one of the longest and most adventuresome expeditions.

The Chibcha and Lake Guatavita

Hohermuth and some 510 troops set out in June of 1535. When they returned, 36 months later, in May of 1538, only some 200 were still alive, among them Hohermuth, but he was seriously ill. They had covered some 2500 kilometres each way and had literally walked almost straight south-west, along the eastern slope of the Cordillera Oriental almost to the source of the Guabero river. They came apparently within 70 kilometres of the easternmost of the Chibcha villages. It meant that they had missed El Dorado as well as Eldorado, the fabulous golden man and the golden country, or paradise, by fluke and by a small distance.

It was really the country of the Muisca, no larger than 300 × 150 kilometres, with probably no more than a few thousand people. The Muisca – they became Chibcha due to a misunderstanding or simply because of mispronunciation – apparently also owned a copper mine called Moniquará, a few days' travel from Zipaquirá. We know almost nothing of this mine, nor of the origin of the gold the Chibcha used other than that the gold supposedly came mostly from northern Equador. The main settlements of the Chibcha were Bacatá (which later became Bogotá), Tunga and Sogomoso.

The origin of the legend of the golden man came apparently from a

Blowing gold dust on to the oiled body of a Chibcha king prior to his diving into Lake Guatavita, near today's Bogota. This practice gave rise to the legend of El Dorado. Source: Johan Ludwig Gottfriedt, *Newe Welt und Americanische Historien*, first published in 1631 by Matthaeus Merian dA and republished in 1980 as Stuttgarter Faksimile Edition im Fackelverlag, Stuttgart, Germany.

religious and political ceremony which the Chibcha enacted whenever a new chief took over. It took place on a small lake, Lake Guatavita, near Bogotá. In the depth of the lake lived – so legend had it – a god whose approval each new king or chieftain had to obtain. People from all over the area gathered on the shores of the lake, with golden and other gifts, while the chieftain had himself covered with a sticky vegetable oil on to which was then blown gold dust which stuck to him, until he was totally covered – the golden man. He was then rowed to the middle of the lake where he dived into the waters to wash the gold off. The crowd also threw many gifts into the lake and legend had it that untold quantities of gold had settled on the bottom. But as yet the Germans and the Spaniards did not reach the Chibcha, did not see the golden man and did not look at Lake Guatavita.

Hohermuth's troops, instead of turning west and up into the mountains, continued further south, until eventually, two years out of Coro, they reached the Caquetá river, near what is today Mocoa. Unbeknownst to them, they were almost at the equator and almost at the borders of the Inca empire. It was August 1537. By then, Pizarro and his army had already taken Cajamarca and Cuzco, and had founded Lima.

The Germans and the Spaniards in the wilderness of southern Colombia were now up against formidable and, as it turned out, unsurmountable obstacles. They could not cross the river – and the way up into the Cordillera turned out to be a nightmare. The rough country finally became too rough, and the natives too hostile. After one of the leaders of the group was killed, the rest prevailed upon Hohermuth to turn back. By May 1538, the remaining, now almost naked troops reached Coro, totally exhausted and with but a few kilos of gold. It had all been in vain.[6]

Or maybe it had not? Back in December 1537, Federmann had left Coro on yet another expedition. He followed Hohermuth's route but eventually went higher into the mountains. When he reached the Ariari river he turned west, and up and into the Cordillera. And he and his troops finally achieved what they and others had so often tried to do – to find the way across the Cordillera Oriental on to the Colombian highland.

Jimenez de Quesada and Antonio de Berrío

And suddenly Federmann and his men were in the land of the Chibcha, and found that other Spaniards had also just arrived – Gonzalo Jimenez de Quesada and his army. Back in 1535, Jimenez de Quesada had signed up as second man to Pedro Fernandez de Lugo who had obtained from the crown the governorship of the area around Santa Marta, founded in 1501 and by 1535 a crucial transhipment port in northern Colombia.

Considerable quantities of gold and copper had been found near Santa Marta since the foundation of the city; this, and the stories and rumours being picked up, were inducement enough for an early expedition into the interior.

By April 1536, Fernandez de Lugo appointed Jimenez de Quesada general of the expedition that now departed. More than 1000 men partly infantry, partly cavalry, were to march inland, more or less southward, parallel to the Magdalena river, while a separate force in six brigantines was to sail up the river as far as possible.

Bad luck, fever, other illnesses, poor food and much mishap, among which the loss of four brigantines, reduced the group at the rate of one man every two kilometres.

After four months they reached the trading centre of Tamalameque, about a third of the way to Bogotá. And they kept encountering salt and gold – and always the reference to more gold further on. However, the troops were disappointed and exhausted and demanded to return. Jimenez de Quesada let them return; he himself pushed on, with only 200 men.

Eleven months after they had left Santa Marta they reached Chibcha country. The natives showed them emeralds, and gold, and gave them potatoes to eat. And then they were at the salt mountain, or salt mine, Zipaquirá. The golden man, El Dorado, the great chief of the Chibchas, was really a salt man. The mine is still being exploited today.

Queseda's forces moved on. In one of the battles they fought as they advanced, the king of the local natives – one of the Chibcha chieftains – got away. It was in his territory – near today's Bogotá and south of Zipaquirá – that they got their first more or less sizeable treasure, some 600 pesos of gold[7] and 145 emeralds.

Jimenez de Quesada eventually reached Tunga and Sogamoso, both north-east of Bogotá, which were run by Chibcha chiefs and where he found considerable quantities of gold and emeralds, enough to cause him to claim that he had found a second Peru.

Yet the Spaniards continued to hear of El Dorado, the golden man, and the search continued. Now they were finally taken to Guatavita, a crater-like lake near Bogotá. But they found no treasures comparable to Peru. And when the soldiers insisted on getting their share of the total so far collected, much turned out to be copper sheet with a thin outer layer of gold. Many of the Chibcha figures were of thin gold sheet, hammered and formed; most were thus hollow and consisted of very little gold. The share for each of the men was small – 500 pesos, or about 3.5 kilograms, for a man on foot; a horseman got double the quantity.

After five months of warfare, much marching back and forth, battles and plunder, the Chibcha civilization was totally destroyed. Their wooden houses, their cities and towns, were burned. Their graves, and the mummies in them, were plundered. As a result, we know today almost nothing of them. The Chibcha civilization was apparently not on the same level as those of the Incas and Aztecs; still, they deserved a better lot.[8]

The Spaniards had no sooner distributed the loot than – coincidences do happen – another army of Spaniards arrived: that of Sebastián Benalcázar.

Benalcázar had been with Cortés in Mexico, had then co-founded the city of León in Nicaragua and had participated with Pizarro in the conquest of Peru. In 1536, he had caught the bug of El Dorado. In 1537 he founded the cities of Popayán and Cali, in today's Colombia. Then he turned south and encountered Jimenez de Quesada who was on his way north again.

And now a third army arrived – that of Nikolaus Federmann, the German, with his troops clad in animal skins, and after a march of 2000 kilometres through jungle and sierra.

It must have been an extraordinary spectacle – three European armies of

160, 189 and 230 men, meeting on a highland plateau in central Colombia. All had left in search of El Dorado. Now they were all there, in Chibcha country. Yet they did not know that they had found it. The legend was to continue, and also the search, for another 200 years, if not more.

Their meeting was surprisingly friendly, in part, no doubt, because they had all found very little and had no immediate dispute over who got what. They decided to found Bogotá, on 29 April 1539, and then to proceed north, to Santa Marta. From there they sailed to Spain to submit their separate claims to the Council of the Indies.

Nikolaus Federmann was to die in 1541 in Madrid, a prisoner of the Welsers and of the crown, as he was about to be tried. The Welsers had accused him of having absconded with gold and emeralds. Benalcázar returned to Bogotá, appointed governor of all land between Quito and Cartagena. But he had a rival – usurper of his position – imprisoned and killed for which act he, in turn, was imprisoned. He died in chains in a Cartagena dungeon in 1551.

Meanwhile, in 1545, the last of the Welser captains, Philipp von Hutten, and Bartolomew Welser himself, had their heads chopped off somewhere in the jungle near the town of Coro. With them ended the rule or participation of the Welsers in Venezuela and the Germans in South America.[9]

It was Jimenez de Quesada who continued the search for El Dorado. He had survived, had pressed his claims at the court, and had eventually settled in Bogotá. Now, in 1569, at age 70, Gonzalo Jimenez de Quesada was appointed governor of the province of Eldorado. He set out on yet another expedition, with 400 men; only 64 returned in 1572. The old man then left his memoirs, his dreams, his debts and his governorship to his niece and her husband, Antonio de Berrío y Oruña.

De Berrío and Doña María travelled to Bogotá in 1580 to commence 12 agonizing years and three painful and fruitless trips east, across the Andes, across the Venezuelan plains and up and down the Orinoco and several of its tributaries. De Berrío found no or very little gold. But he spent a fortune to finance his trips.

And the Guianas, meaning the province of Venezuela that bears the name (south and south-east of the Orinoco), the Republic of Guyana, the Republic of Surinam, the French colony of Guiana and the Brazilian state of Amapá (topographically all bunched together as the Guiana highlands) are all major mining countries. Large deposits of iron and manganese ore, bauxite, placer gold, deep rock gold, cassiterite and many other minerals are there. Some are already being exploited, others are known to exist, still others are no doubt still awaiting discovery.

There is even a Colombian province of Guainia, in the easternmost bulge of the country, the one that borders on Venezuela, where the countryside must be extremely wild, and where innumerable rivers flow east, towards the

Orinoco. This region no doubt also got its name from the old legends. Over the centuries a trickle of prospectors must again and again have crisscrossed the area, and must have panned for gold. But fame the area never achieved. It is only now, towards the end of the twentieth century that, with government help, the area is to be opened up.[10] Much gold will no doubt be found and, judging by what is happening in the adjoining region of Brazil, also cassiterite.

But we must return to earlier days, to the period 1690–1810.

It was during the third of de Berrío's trips and while he was settling the island of Trinidad, in 1595, that Sir Walter Raleigh arrived. Raleigh burned the settlement, took possession of the island for England, and imprisoned de Berrío. Released a few months later, de Berrío finally died in 1597 at age 80, a bitter and disappointed man.

Sir Walter Raleigh and the Guianas

Before de Berrío died, Raleigh had skilfully wormed out of him whatever information – and dreams – the old man had of Eldorado.

Raleigh was 43 years old when he began his search for Eldorado. He had already spent many years and six expeditions, since 1585, to settle Virginia. Now he was to pursue the new dream for another 23 years before King James I was to chop off his head, in 1618, as a sop to the king of Spain.

Raleigh was said to have had the ambition[11] to set up a great English empire in South America. He wanted to destroy the Spanish might. But he did not succeed. Feeble only were British attempts in subsequent centuries, leading eventually to wresting Trinidad from Spain in 1796, and establishing British Guyana in 1834, besides, or course, Jamaica, which the pirates had first occupied and which had become British in 1655 when an army under General Robert Venables simply occupied the island.

Raleigh, though unsuccessful in finding Eldorado and gold, did with his reports excite the imagination of the English, the Dutch and the French more than the Spaniards ever had, and that also eventually led to the Dutch and the French establishing themselves in the Guianas.

Gold and platinum in the Chocó

When we look at a map of South America, the north-westernmost corner is the Chocó. It is a small area, south of Panama, north of the Calima river, west of Medellin and north of Cali. Even today it remains a wild, almost impenetrable country. The climate is forbidding, the isolation almost complete. Yet between 1680 and 1810, the Chocó was one of the prime gold mining centres of the world.

The first white men to enter the Chocó must have been Pascual de Andagoya and his men when they sailed down the west coast of South

America, in 1522. Subsequently, Pizarro arrived in 1524. But the natives were extremely hostile and did not allow penetration, which was only possible up the San Juan river from the Pacific Ocean, or up the Atrato river from the Caribbean Sea. The hostility of the natives was, in fact, so fierce, the weather so inclement, disease so rampant, and nature so forbidding that 200 years were to pass before, in about 1680, white prospectors finally entered in force. By 1684, most of them were dead, as a result of native resistance. It took another two years and determined efforts by the Spanish authorities before the region was finally at peace.

Gold was all over the area, in the rapid rivers, the creeks, the ravines, the dry creek beds. It was all placers, alluvial deposits. Most gold was produced by means of ground sluicing. And the principal raw material for sluicing, water, was almost always available in abundance.[12]

The total mined during the 130 years or so that the boom lasted has been estimated at anywhere between 40 million silver pesos and double that amount. Based on the tabulations made from the tax records, it would be 40 million pesos.[13] But as we know from many countries of the Spanish colonial empire, probably only half that found was officially declared – and a similar quantity was again stolen, or not reported. For instance, in 1789 Francisco Silvestre, former governor of the province of Antioch and secretary to the Viceroyalty of New Granada, estimated that more than half of the total production had probably been smuggled out.[14] If one thus doubles the tax tabulation of 40 milllion pesos, about 80 million silver pesos would have been the quantity of gold produced in the period. Eighty million pesos would weigh about 360 tonnes.

As already said previously, values in the olden days were iffy. The silver peso was said to have been equal to a silver dollar in around the year 1800.[15]

The prospectors also panned platinum from the same rivers; it usually appeared together with gold. They called it 'white gold' and thought it was a type of immature gold which had not been long enough in the ground to mature and turn yellow. For a while they put it back into the ground, so that it could mature more.

From about 1720, white gold was considered fraudulent, and he who brought it mixed with real gold was fined. Spaniards caught had to pay a fine of 2000 pesos, about 10 kilograms; others, of inferior caste, got 200 lashes.[16]

Platinum was probably refined for the first time in Bogotá, at the Casa de Moneda, in 1774; from then on it came to be used in Europe. The Spaniards had finally come to realize that platinum was as malleable and as ductile as gold – and cheaper in those years, 1765–95, when it was still considered an impurity.

Antonio de Ulloa and Jorge Juan, Spanish civil servants who in 1735 accompanied the French count and scientist La Condamine on an expedition to Equador and Colombia to measure distances at he equator (but who also acted as secret agents for the Spanish crown to report on general conditions and the reasons for drastic declines in mining production in the Spanish colonies), are credited with the discovery of platinum, its availability and possible uses.[17]

Towards the year 1800, gold and platinum production in the Chocó eased off, and declined rapidly after the turn of the century. The industry had been based on slavery, first of the natives and then of imported blacks. By 1760, practically all labour was black. But when the international slave trade declined, it naturally reduced the influx of new labour into the Chocó and affected mining activity. New slaves could not be bought, and manumission, death and flight caused the numbers to decline. And as the business became less profitable, for lack of cheap labour, the operators withdrew. Eventually a quite primitive population of *mestizos* was left. They continue, to this day, with steady yet relatively small placer operations.

The Chocó must have been a wild place, in those years.

In 1660, the native population was estimated, by Jesuit missionaries, at 60 000. But by then, several expeditions had already explored the area and as a result the native population had probably already been reduced. By 1763, the number was down to 4732, although this census figure did not include those tribes still not conquered.

The number of slaves went as high as 7088 in 1782, but again, this did not include those who had run away and were living in their own communities deep in the mountains and ravines. The number of officially registered and counted freedmen reached 15 184 in 1808, the last time such a census was made.

The whites in the Chocó never numbered more than a few hundred.[18]

Heavy rainfall, jungle, swamps, humidity and forbidding mountain ranges are typical of the Chocó. It seems to rain all the time, as much as 350 inches on average a year. The humidity must be close to 100% all year around but the average yearly temperature is only 82 degrees Fahrenheit.

Venezuela's El Callao mine – the real Eldorado?

The El Callao gold deposit, a real deep rock mine, lies close to the Yuruari river, south-east of Puerto Bolivar and 100 kilometres from the border with the Republic of Guyana. For a while, between 1842 and 1912, the mine was famous for its production. It paid fabulous dividends. The town of El Callao rose around, or rather, above it.

Auriferous sands were reported first in about 1840, no doubt as a result of casual visits by prospectors. But old Spanish workings were also discovered, long abandoned, probably during the years of the war for independence (1808–26) when Spanish citizens fled the country. Records of what the Spaniards did and worked have not been found.

Back in 1842 it had been a Brazilian, one Pedro Joaquim Ayares, who had appeared at Ciudad Bolívar with gold samples he claimed came from the Yuruari river. But only in 1849 was credence given to the stories. A Frenchman, Dr Luis Plassard, again brought stories of gold finds. But it was then still placer gold.

The first washings began in about 1849 and the old mine workings – a real lode mine, were probably opened up again. It is regrettable that we have no reports on what had gone on previously, and what the first post-independence prospectors found.

The statistics show that in 1871 about 100 kilograms of gold were produced. In 1884 it had gone up to 5 tonnes. By 1892, production was back to 1 tonne for the year, and by 1912 had ceased altogether.[19]

What was El Callao? Lucien Morisse, a French medical doctor who must, however, also have had a more than cursory interest in mining in the backlands of Venezuela as he eventually mentions his 'dredge' on one of the rivers near El Callao, visited the area in about 1900 and wrote a book called *Excursion to El Callao* which was published in Paris in 1904. Most of our information comes from his book.

It was a gneiss deposit with a thick quartz vein that ran from the apparently rather visible and rich outcroppings on the surface – which no doubt were what had originally drawn the attention of the prospectors to the deposit – at an angle of about 35 degrees into the ground, all the way down to 250 metres. There the vein apparently levelled off and then continued more or less horizontally for a while.

In about 1870, a first mill was erected. It came to crush as much as 74 000 tonnes, in 1873.[20] In the 1880s, the developers began planning a new mill and it started to function in 1887. Curiously, or ironically, that was also the year they lost the vein.

The real impetus for the workings had originally come from a Corsican trader and mule train operator, Antonio Liccioni, from Ciudad Bolívar who, in about 1860, crisscrossed the jungle area south of the Orinoco. Liccioni eventually registered some claims and, in 1870, with 11 partners formed Compañia El Callao, to open up and operate again the old Spanish underground workings. Several shafts were sunk and tunnels were driven to follow the quartz vein. They did find gold. The grade soon ran between 30 and 250 grammes per tonne. But the vein was irregular, water was a major problem, and working the deposit without a mill appeared impossible. They had to borrow to get materials and supplies. But eventually they managed to sell some shares in Paris.

Once the mill was in place, larger quantities could be processed. Now the grade went up to 400 grammes per tonne.[21] By 1875, El Callao paid its first dividend – double the original investment. In 1876, the dividend amounted to five times the investment[22] and in 1886, it amounted to 34 times the original capital.[23]

They paid dividends uninterruptedly for 11 years; sometimes every month. By the mid-1880s, Liccioni was a millionaire, with properties in Paris and elsewhere. He was even decorated with the Légion d'honneur.[24]

For a while, during the rich bonanza years, wages were paid with quartz – 'take out at night what you can carry'. At times the walls of the tunnels were encrusted with thick sheets of pure gold.[25]

But in 1887 the vein suddenly petered out. Just like that! No more gold content, no more quartz, no more vein. It took over a decade to work the tailings down, before they gave up.

But Morisse still saw the glittering quartz vein and saw the gold incrustations in the pillars – ready to be scratched out with a knife.[26] Charming stories were remembered about the ground and the adobe huts around the old mill still being full of gold dust and even nuggets which were recovered as they processed the loose gravel and even entire adobe huts throught the mill.[27]

Morisse also still saw several of the old Spanish surface workings as well as alluvial placers worked until the independence days.

Later, after they had lost the vein, and after they had also shaved the underground pillars down to less than safe thickness, the tailings were still running about 30–60 grammes per tonne. But in 1894, the mine collapsed. They had taken out too much, and too many of the pillars, without replacing them with timber.

How much did the old placers yield, and how much the El Callao mine? We don't know. The statistics are incomplete. And there was the theft factor. What the workforce took out, in the form of pay which was never registered, can simply not even be estimated. But Morisse was certain that El Callao was the original Eldorado. It was not even the only mine in the area. Others were no doubt worked in colonial times, and maybe even in pre-Spanish times.

Sporadic work at the El Callao mine and other nearby placers has continued throughout the decades, with gold production fluctuating widely, all the way from zero to 2.5 tonnes per year.

The Caucá valley and the Buriticá mine

The Colombian department of Antioch lies to the east of the Chocó. The Caucá river runs right through it, the full length, northward. Its two principal cities are Medellín and Antióquia.

As in the Chocó, the natives who inhabited the Caucá valley proved formidable opponents to Spanish penetration when the conquistadors began to move up the river from the Caribbean, after some initial attempts to do so from the town of Buenaventura, on the Pacific coast, which they had founded in 1535.

Here too the Spanish went for gold. And they found it, in substantial quantities, and over the centuries. The Caucá valley, or even the entire province of Antioch, should probably, and with considerable justification, be called one of the principal gold producing regions of the Americas.

The Spaniards entered the area from the Caribbean, up the Caucá river, across the Andes from Buenaventura and Cali (founded in 1537) and from the highlands of Equador and the Bogotá plateau. One of Benalcázar's lieutenants, Jorge Robledo, is credited with conquering and then settling the Caucá valley. The battles with the natives were bloody; most of them were ferocious cannibals. Cieza de Leon, the chronicler who accompanied Robledo, often saw piles of human skulls outside native huts, which the Spaniards thought were butchers' shops.[28]

But the Spaniards did not find a lot of gold. No golden cities, no men clad in golden armour. But when they were led to burial mounds, they found many golden artefacts buried with the dead. Systematic plundering of graves followed.[29]

The natives also led them to a real gold mine, the lode mine of Buriticá. This must already have been a fully-fledged industrial centre in pre-Columbian days, and it may well have been one of the few lode mines in the area. The Spaniards found crucibles, braziers and scales. The natives must have melted placer gold (rather pure dust and nuggets) and smelted gold-containing ore.

There the Spaniards founded the city of Antióquia, or Antioch, in 1541, and then proceeded to exploit the mine for the next 80 years. Many placers along the Caucá river and at nearby streams were also worked and yielded substantial quantities of gold as did, between 1576 and 1685, the Caceres placers, near the city of Caceres and, from 1602 to 1688, the Zaragoza placers, near the city of Zaragoza.[30]

The placer boom in the Caucá valley lasted well right through the eighteenth century, practically until the advent of the independence movement early in the nineteenth century.

Few statistics are available as to what was produced. Most of the gold that was exported from Cartagena in the early years of the conquest supposedly came from the Caucá placers and the Buriticá mine. Statistics exist for the period 1670–1807, but they are incomplete as a number of years are missing. Some 6 million gold pesos were registered during the years for which details are available. If we extrapolate similar yearly quantities for the missing years, the total for the entire period would leave us with something like 9 million pesos (of gold) or, at 4.55 grams per peso, about 41 tonnes for the period. As the theft or contraband factor no doubt also applied here, total production could well have amounted to something like 80 to 100 tonnes over the period of about 138 years, or about 600–700 kilograms per year. More meaningful perhaps is that during the last 30 years before the independence wars, the part that was registered reached about 1 tonne per year, and when we include the smuggled quantities, it could be double that, i.e. about 2 tonnes a year.

Colombia and Venezuela today

The iron ore deposits of eastern Venezuela must have been known for a long time, and must have been looked at again and again; and some raw iron may well have already been made early in the eighteenth century. In 1782, a progressive and perhaps even visionary viceroy not only informed Madrid of the rich deposits but even suggested that such ore be shipped to Spain, maybe to replace Swedish ore. But the Spanish crown forbade further investigations.[31]

It had, in fact, been as early as 1743 that Capuchin monks had begun working some iron ore in the Santa Rosa hills, south-west of Upata. One of Simon Bolívar's decrees declared all mines property of the nation, in 1829. But it took until 1883 before there was real interest in iron ore mining and the first exploitation concessions were granted for the Orinoco delta. But even those were apparently only insignificant ventures.

Meaningful interest in large scale industrial exploitation of iron ore only came in the 1950s when a subsidiary of Bethlehem Steel began opening up the Cerro El Pao mines, and United States Steel opened up the Cerro La Parrida deposit. In 1975, the government of Venezuela, 'exercising its sovereign rights', nationalized the deposits.[32]

Already in 1960, a government owned company, Corporación Venezolana de Guayana (CVG), had been formed, to take charge of and develop major ore deposits in the eastern part of the province of Bolívar. Much of the iron ore mining is concentrated around Cerro Bolívar, some 30 miles south of Ciudad Bolívar. But the mineral wealth extends all across the province, north-east to the Orinoco delta, east to the border with Guyana (including the gold bearing area around El Callao) and south to the border with the Brazilian state of Roraima.

The area not only yields iron ore. CVG and other interests have also set up (or are in the process of building) several large steel works. The Caroni river has been dammed; the resulting cheap hydroelectric power supply has allowed the setting up of several aluminium smelters to process the large bauxite supplies also found in the area. There are also ferroalloy plants.

Venezuela has thus become, in little more than 30 years, a world-class power house in iron ore and steel production, bauxite and alumina, as well as aluminium.

And gold production has also not lagged. Though still small, at 14.2 tonnes in 1990, it had grown spectacularly from only about 1 tonne in 1980, and may well continue to grow in the future.

Colombia, however, remains a land of promise unfulfilled. Small quantities of copper and zinc have been produced, but only sporadically over the last 10 years; more recently some nickel has also been produced. There is also silver, something like 6 to 10 tonnes a year. Gold production doubled between 1980 and 1989 and will, no doubt, grow further. In 1990, with 32.5 tonnes,

Colombia was the second largest gold producer in Latin America. In 1991, production dropped back to 31.6 tonnes.

We can imagine that each country's current annual gold production is probably equal to if not a multiple of whatever treasure the Spaniards found and took 500 years ago.

Emeralds forever

Emerald mining is not really part of metal mining, because these semi-precious stones are not metallic. Yet the story of the Colombian emeralds should be told.

Emeralds are the most prized of the beryl ores – coloured by just a touch of chrome which gives them the green hue. Everybody cherished them, the Romans, the Greeks, the Egyptians as well as the rulers of India. The earliest stones came apparently from Cleopatra's mines in Egypt, near the Red Sea.

And then the Spaniards found them in Colombia. But it was not really the Spaniards who found them. The natives, the inhabitants of the Bogotá plateau where the mine is located, the Chibcha, must have dug them up for centuries if not for a millennium before the arrival of the conquistadors. Stones which can only have come from the Colombian deposit had fallen into Spanish hands already in the Caribbean, and then in Mexico and Central America. They must have been taken there by long distance trade over considerable spans of time.

It took the Spaniards a while to get there. By the time they reached a place called Chivor, in 1537, some 65 miles north of Bogotá, they had already robbed the native villages through which they had passed and had amassed a treasure of some 7000 emeralds. The mine itself was only found, or revealed to the Spaniards, in 1558.

Since then, the Muzo mine has produced most of the best and largest emeralds the world has ever known. Something like $500 million worth of emeralds are mined each year. But more than 50% are supposedly smuggled out of the country. The official 1988 Colombian statistics show exports of emeralds to the United States totalling $11.5 million. Yet US Customs show that $42 million worth were declared as imported that year.[33]

It is an open pit mine. The emeralds are encrusted in black shale powder. It is simply a primitive surface bulldozer operation, scraping the black slate and exposing hydrothermal veins of calcite that carry the gems.

One of the more charming stories about these emeralds goes back to the early years of the conquest. It is about a Dominican priest by the name of Reginaldo de Pedraza who, in 1531, accompanied Pizarro on his trip down the Pacific coast of Colombia towards Peru. In Equador, the Spaniards found, or expropriated, or obtained through barter, quite a number of emeralds, some as large as dove eggs. The padre told them that the difference between real and false emeralds was that, when beating them, the real ones would not crack. And, as the Spaniards wanted to be sure the stones they had were real emeralds, they hit them all. The padre is reported to have become rather rich, later on, when he sold the broken pieces in Panama.[34]

Equador – an unlucky fault, or blanket

Equador seems to be the least fortunate of the Andean countries in terms of metal mining. Though the same mountain chain runs through all the countries, Colombia, Equador, Peru, Bolivia and Chile, and even Central America, Mexico, the United States and Canada, Equador appears to be the only country without major mineral deposits.[35]

The explanation seems to be that a blanket of recent volcanic rock covers most of the mountain area of the country, burying the potentially minerals-bearing strata.

Some mining there has of course been. An old, old gold mine at Portovelo has been worked, off and on, since Inca times. Some copper deposits have been worked, also sulphur, coal, iron and manganese ore, as well as silver, lead and zinc. This shows clearly that the rock formations below the country seem to be basically no different from those of Peru. And yet metal mining never amounted to much in Equador, except maybe for the years of World War II when metallic products, essentially gold concentrates, copper concentrates and copper bars, accounted, for a few years, for meaningful percentages of export values. Whereas in 1930 mining products had accounted for 9% of all exports, the percentage jumped up to 24% in both 1940 and 1941 to then fall back again to 12% by 1945, 5% in 1950 and less than 1% in 1957.

We shall now turn to Peru, where the Spaniards arrived next, back in 1530, and which then already was and continues to be today a mining country par excellence.

Notes

1. Mazur, 'Visions of the Alto Magdalena'.
2. Ibid.
3. Dalton, *Venezuela*, page 66.
4. Konetzke *Fischer Weltgeschichte*, pages 302–3.
5. Though it was a German undertaking, the company was mixed.
6. The stories of the Germans marching up and down the rivers and plains and cordilleras of Colombia are wonderfully told by Victor von Hagen, *Auf der Suche nach dem Goldenen Mann*.
7. Equal to about 2.7 kilograms.

8. Von Hagen, page 139. In Tunga alone, they had found or taken 136 500 pesos of gold, about 628 kilograms, and 230 emeralds. Victor von Hagen published several books on pre-Columbian civilizations and the colonial history of Peru, Columbia, Venezuela, All are very readable and immensely informative. See bibliography.
9. Not quite. The Fuggers also tried, in about 1531 when Charles V gave them most of southern South America, to explore and exploit, but the big Fugger fleet disintegrated in the Atlantic in 1532, and the project was abandoned (Ogger, *Kauf Dir einen Kaiser*, page 298). The Germans – to be precise, the Brandenburgers – tried it once more, in 1690 in St Thomas, as we already saw.
10. Rutledge, 'Columbia plans gold revival'.
11. Hanson, *South from the Spanish Main*, page 236.
12. Sharp, page 48. Part of each year, however, the rains were too heavy and the rivers too full. William Frederick Sharp's book, *Slavery on the Spanish Frontier: The Columbian Chocó, 1680–1810*, is an extremely valuable contribution to the history of mining in Colombia, especially during the colonial period. It is also a remarkable example of how historical research is done, and how surprising results come about. Mr Sharp went to the Chocó as a Peace Corps worker and began his research then and there.
13. Ibid, page 73.
14. Ibid, page 71.
15. Ibid, page 10 of the preface.
16. Ibid, page 51.
17. Prieto, *A Mineração e o Novo Mundo*, page 61.
18. Sharp, page 199.
19. Dalton, page 216.
20. Ibid.
21. Morisse, *Excursion a El Dorado – El Callao*, page 123.
22. Ibid, page 124.
23. Ibid, page 128.
24. Ibid, page 126.
25. Ibid, page 124.
26. Ibid.
27. Ibid, page 143.
28. Twinam, *Miners, Merchants and Farmers in Colonial Colombia*, page 15.
29. Ibid, page 16.
30. Ibid, pages 16–17.
31. Konetzke, *Fischer Weltgeschichte*, page 300.
32. Pamphlet with general information put out by Corporación Venezolana de Guayana.
33. Star, Fred, 'The timeless mystique of emeralds', *National Geographic Magazine*, July 1990.
34. Von Hagen, page 127.
35. Linke, *Ecuador*, page 142.

CHAPTER 6

Peru

Travellers who go today from Panama to Peru cover the distance by aeroplane in a few hours, and almost in a straight line. They will probably not see much of the coast of Colombia, but the flight path will go across the Andes of Equador and northern Peru. When the Emperor Charles V asked what Peru looked like, Francisco Pizarro is said to have crumpled a sheet of paper, explaining to his majesty that this was how the country looked. The story is probably apocryphal (it is also attributed to Hernán Cortés and Mexico), but it could be true. The mountain ranges of Equador and Peru, as well as Bolivia, Chile and even Mexico, look very much like crumpled paper. They appear bare, stark and forbidding.

The rumours of gold 'to the south', of a golden empire, and of easy pickings and wealth, led the Spanish conquistadors to set out for these regions soon after Vasco Nuñez de Balboa had, in 1513, discovered the Pacific Ocean. Panama was founded in 1519; from there, Pascual de Andagoya sailed south in 1522.

But the discovery and conquest of Peru and the Inca empire are forever associated with Francisco Pizarro, his brothers and other relatives, and a few hundred Spaniards. Pizarro made a first trip in 1524–5, a second in 1526–7, and then set out, in December of 1530, on the third and definitive voyage which led to the breathtaking events of Cajamarca and Cuzco. History books and historical novels have again and again retold these stories.

The Peru they discovered, conquered and then settled has also become, like most of the Andean nations, an area where mining has played and continues to play an important role in the life of the nation. No other

Table 6.1. Schematic dating of the pre-Columbian civilizations in Peru

Civilization	Dates
Inca	AD 1438–1532
Chimu Chancay	AD 1300–1438
Tiahuanaco Ica	AD 1000–1300
Mochica Nazca Recuay	AD 400–1000
Paracas	400 BC–AD 400
Chavin	1200 BC–400 BC

organized activity has contributed so much to wealth and development as mining. Greed for its supposed mining wealth brought the first Europeans to its shore, and still today much of the hope of the country is focused on mining, present and future, sustaining and improving the country's growth.

We know very little of mining in Peru prior to the year 1532, the year the Spaniards occupied it. But we can safely assume that placer gold from the rivers and outcroppings of silver and copper were the first metals discovered and used, thousands of years ago. Found in a relatively pure state, they were easy to work, principally by hammering, into desired shapes of ornaments and idols for worship. The effect of fire on metals was probably discovered accidentally, at an indeterminable date. Copper was worked mainly for its usefulness for weapons.

Although many of the early civilizations of Peru were familiar with, and used, metals, the Mochicas, or Moches, in the period AD 100–1000, and the Chimu, in the period AD 1200–1470, appear to have been the undisputed masters of early metal working. They knew how to smelt and alloy metals and, judging by descriptions the first Spaniards left and the samples which survived, made jewellery and adornments of intricate design and craftsmanship. This excellency declined after the conquest and the destruction of the Chimu empire by the Incas in around 1470. Its capital, Chanchan, became a ghost town soon after. Many of the artisans moved, voluntarily or forcibly, to Cuzco, the capital city of the Inca empire, to henceforth work for new masters.

The Spanish conquest

In Peru, as elsewhere in the Americas, nobody counted heads in those early years. Yet several of the early chroniclers guessed at the population numbers. They came up with a figure of maybe 30 million, or maybe 15.[1] Later estimates reduced the number to 10, or 8, or even only 3 million people at the time of the conquest, based on modern concepts of food production.[2] But we

Spanish conquistadors arguing over the division of spoils, about 1540. Source: Guaman Poma de Ayala, one of the early chroniclers.

do know that by 1796 a census showed 1 076 000 inhabitants.[3] By then, presumably, the native population had already begun to recover once more. The point is that in Peru, as elsewhere, the conquest led at first to a drastic decimation of the native population.

It still dazzles us how a mere handful of people, about 150 in all, with very few horses, faced up to and overwhelmed a native army of probably some 20 000 warriors in the town of Cajamarca, and then went on to subjugate the entire empire.

When the Spaniards had the Inca leader as their prisoner, following their surprise attack, they were stunned when Atahualpa offered to fill several rooms with gold and silver artefacts as a ransom. The Inca leader not only promised, he fulfilled – only to be garroted, for political reasons.

The Spaniards then moved on to Cuzco and, in 1533, also plundered that city.

In terms of weight, the loot from Cajamarca came to about 5 tonnes of gold and about 10 tonnes of silver, and from Cuzco to about 3 tonnes of gold and some 30 tonnes of silver.

The Spaniards next turned to the sources of the gold and the silver. Where were the mines that had given such wealth? It was not easy to find them. Under the Incas, little systematic mining had taken place as gold and silver had no monetary value, were owned by the Incas and were only needed to become decorative pieces. Thus, the Spaniards found few pre-conquest mines – and the history of mining in Peru really covers what happened in colonial and republican times.

Under Spanish rule, several fundamental changes occurred. Now, suddenly, gold and silver had value, were coveted, meant wealth. Practically all the masterfully worked and beautiful pieces of art disappeared into the furnaces of the conquerors, to be converted into bars and coins. The loot made a few, quite a few, very very rich. And it gave much sudden wealth to the Spanish crown. Very little eventually returned as adornments to the churches of Peru, Bolivia, Equador and Colombia.

Colonial times

The period 1548–78 is of singular importance in the history of mining in Peru. The civil wars among the conquerors – to which most of the Pizarros and many of the first conquistadors fell victim – were finally over, and some peace and order began with the first royal administrators. Into this period falls the discovery of Potosí and the first large scale and very profitable exploitation of the mining wealth of the country. It was a major historical coincidence that the Santa Barbara mine near Huancavelica – then and until recently the only significant mercury mine on the entire South American continent – was discovered at the same time, because towards the end of this period, when the high grade surface deposits of Potosí gave way to lower grade ore, the mercury from Santa Barbara came as a godsend. It allowed a second and sustained silver boom through large scale application of mercury in amalgamation. It was the following continuous and massive flow of silver from South America, mainly from Potosí, which caused the many changes in Europe, in finances, commerce and industry, mentioned earlier.

In Peru, as in Mexico, colonial mining was not based on scientific principles. Technical progress was slow. Medieval methods were perpetuated. Exploration, as we understand it today, did not exist and thus few new mines were discovered, Potosí and Santa Barbara being notable exceptions. The Spaniards in general continued to work the old native, pre-Inca and Inca veins. Yet knowledge of many mines worked by the Incas must have been lost as their location was hidden from the Spaniards; little was revealed even under torture.

As the known mines were worked out, or work reached lower and lower levels where, with the technology then available, exploitation could not continue, and no new mines of importance were discovered, production and the flow of wealth slowly declined.

The Santa Barbara mercury mine

The town of Huancavelica lies at 3676 metres above sea level. Its population is today probably 20 000. The town owes its origins to the mine which lies – almost literally – right above it. Although the mine was discovered in 1563, it was only in 1572 that Viceroy Francisco de Toledo founded what he called 'Villa Rica de Oropesa', after the castle where he was born. The name Huancavelica came later and derived from two Quechua words meaning 'stone idol'. The town became a city in 1581.

Huancavelica's fortunes rose and fell with the fortunes of the mine. Much production meant activity and prosperity; suspension of work spelled doom for the city. Although Huancavelica is the capital of the department with the same name, the terminal of a narrow gauge railroad which connects it with Huancayo and also the warehousing and supply centre for a number of mining companies that operate in the region, it has never really 'made' it. No industry has settled here. Nearby Huancayo is bustling, sprawling, modern and ugly, with possibly as many as 250 000 people.[4]

One thousand metres above Huancavelica is the mine. One can see from the town square the ravine up and down which, in colonial days, all the traffic to and from the mine went. Today a dirt road winds for many kilometres up the hill, connecting the town with the mine, or what was the mine. Actually, Huancavelica also lies in a deep ravine, much below the surrounding country. The region is mainly puna; very little vegetation grows anywhere. Agriculture is confined to the bottom of the valleys. Up in the puna one only finds alpaca, sheep and llama herds. The seediness, the slow pace, absence of traffic and the small scale of everything give Huancavelica an air of the past. The future has as yet not arrived and, actually, the present is perhaps not even there. And the past, the days of wealth and glory of the sixteenth, seventeenth and eighteenth centuries, is long gone.

The Romans called it cinnabar; it is an orange-coloured ore which exudes dangerous fumes. Yet in antiquity, the ore was ground to a powder and used by ladies to add colour to their cheeks. This ore contains a metal, mercury.

Some call it quicksilver. It looks shiny, like silver, but it is not. It is the only metal which is liquid. It can be extracted by roasting the ore slowly whereupon the volatile metal escapes into the air as fumes. If the fumes are trapped and allowed to cool and condense, liquid metal forms.

A description of the patio process has already been given (see page 30), and shows how mercury, when mixed with finely ground silver ore and other ingredients, binds the silver in the form of froth or small bubbles which, when roasted, give off into the air the mercury in the form of fumes. Almost pure silver is left behind. The Romans apparently knew the process as Pliny mentioned it in his writings.

During the Middle Ages, knowledge of the process was lost. It was rediscovered in Spain, in the fifteenth century, by German metallurgists. But though the process was available in Europe, it was not applied. It took a shipment of Almaden (Spanish) mercury to Mexico for the process to be applied, by Bartolomew Medina, in 1555.[5]

Almaden mercury was soon also sent to Peru, mainly for Potosí. The first such shipments probably reached Peru in about 1558.[6] Potosí had been discovered in 1545 but the first big bonanza was already over by 1555 and the yields began to decline. No more almost-pure silver to be extracted from the mountain! Mercury was now needed for amalgamation, to extract the silver from the lower grade ores.

And then, suddenly, mercury was discovered in Peru, about halfway between Cuzco and Lima and not far from the road that connects the two cities. Santa Barbara was to produce, with few interruptions, for the next 400 years.

One story is that a Spaniard by the name of Amador Cabrera took part in a religious festivity in the town of Huamanga. It must have been in about 1560. He supposedly left his jewel-studded hat in the care of the son of an Inca noble so as to better amuse himself without the cumbersome headpiece. The boy lost the hat, or maybe it was stolen from him, but the boy's father promised to make up for the loss. And he then led Cabrera to some outcroppings of orange-coloured ore which became the Santa Barbara mine.

We know that the Incas, like the Romans, used the ore as a rouging agent, but no more. It is generally claimed that they did not know that the ore contained mercury and that they did not know of the amalgamation process. But they may, after all, have known. Some chroniclers report that primitive furnaces of pre-conquest origin were found near the mine. They must have used them to roast the ore. As roasting did not change the quality of the rouging agent, the furnaces must have served another purpose. Perhaps to produce mercury?

Cabrera took a good look at the ore, the holes in the ground and the veins, and presented himself, on 1 January 1564, at the town council of Huamanga to file his claim to the mine. He called it 'La Descubridora'. He also delivered a small jug of metallic mercury which he had already produced and he swore that it came from the mine. He was given the rights to the claim. It was later re-baptized 'Todos los Santos', All Saints, and still later 'Santa Barbara'. As soon as Cabrera's claim was announced, others rushed up the hill. Eventually a total of 43 claims were registered. Cabrera and others quickly became rich, and very rich at that.

But the mercury was too important to the crown to be left in the hands of private entrepreneurs. In 1572, Viceroy Toledo expropriated the mines and negotiated work contracts with most of the important miners who then continued exploitation on behalf of the crown, for a fee.

The mine continued to give substantial profits to the operators. They continued to dig, and dig. Actually, the burrowing was done by the natives. They were simply ordered to go and work in the mines. The system was called *mita*. Part of the overall taxation scheme was that each village had to make available, for specific periods of time, a certain number of people. It was, on the part of the Spaniards, simply the continuation of a system which had already existed under the Incas and which in a way was similar to the serfdom of feudal Europe. The American natives, like the serfs, had to work part of the time free of charge for their lord.

They were ordered to go underground, and most of them simply stayed and died there. Food was taken down. They laboured in near obscurity, day-in, day-out, always exposed to and swallowing the dust and the slight mercury fumes. Rules that there should be days of rest were ignored; the concession holders were only interested in getting the maximum quantity of ore out of the mine. The natives worked with very primitive tools and simply followed the colour of the veins, digging ever deeper into the mountain.

Mining technology was still practically non-existent and whatever systems had been developed in Europe were certainly not available to the rather empiric miners in the highlands of Peru. Furthermore, during the first 100 years of exploitation of the Santa Barbara mine, scientific methods could probably be ignored because labour was so cheap. So what if there were occasional cave-ins because of haphazard digging and no safeguards? Many natives died, at times several hundred in one accident. But that did not matter since there were always more of them.

As time went on, Santa Barbara became known as 'the mine of death'. It was not the slave labour system of working permanently underground that gave it this reputation; it was not even the occasional cave-ins, not even the big ones following earthquakes. It was rather that whoever went underground and stayed there for some time inhaled too much mercury bearing dust and became incurably ill. The natives of central Peru came to dread having to go to Santa Barbara.

The mine was described repeatedly as being a real underground city, with roads, squares, chapels and all kinds of service quarters. One of the roads was said to have been wide enough to allow, from time to time, bullfights to take place, by torch light. It must have been a Dantesque spectacle.

The very rich veins were soon worked out and the grades, which at first had run in excess of 10% mercury, declined. But the quantity of metallic mercury produced above ground from ever larger quantities of ore remained reasonably constant, with some fluctuations, until 1786, when a terrible disaster hit the mine. It collapsed. An overambitious administrator by the name of Marroquín had removed many of the supporting pillars of ore which had been left standing during the preceding 230 years of work. Marroquín wanted to raise production, and substantially so. He ignored warnings and proceeded to remove the pillars to process what was indeed mostly high grade ore.

More than 200 people died; Marroquín was imprisoned. Production now declined rather rapidly. Between 1793 and 1796 mine operations were discontinued yet some production was maintained by letting anyone go through the dumps. Since much high grade ore had previously been cast aside, some 1800 tonnes of metal were recovered over the next 10 years.

In 1806, the crown gave orders to destroy the mine, but the orders were ignored. The royal administration now withdrew and the mine was left to haphazard exploitation by small miners. In 1813, it was finally abandoned.

The production between 1565 and 1813 has been tabulated from royal records as totalling 55 000 tonnes of metal. We should suspect that actual production was probably closer to double that figure as much mercury was no doubt stolen and smuggled out.[7]

A second period of production lasted from 1820 to 1863, during which a further 7500 tonnes were taken out. In the latter part of the nineteenth century, the mine changed hands repeatedly but was not worked. It finally became the property of one Eulogio Fernandini who, first in 1915 and then again in about 1940, tried to open the mine up again. It was only after his death that new ideas were tried out. Beginning in 1955, large scale open pit operations were begun and the processing of the remaining low grade ore finally made sense. Once more rather substantial quantities of mercury were extracted, for a while. The story will be told further on.[8]

But back to mining in Peru during the last years of the colonial administration. By the year 1800, the industry had almost totally collapsed. What was left of the viceroyal administration was corrupt and inefficient. New political ideas were spreading, from North America and from France. The Spanish rulers, by now the Bourbons, were busy with European matters. And the ever smaller production of silver as well as the ever higher administrative cost reduced the flow of metal and taxes to a mere trickle. Peru, and South America, had probably already become a burden to the mother country.

Laicacota and the Salcedos

'Laicacota' was originally the name of a hill, but the word really came to be identified with a famous silver mine near Puno. It was also the name of a town, about 2 kilometres north of today's Puno. It was, in a way, the predecessor of Puno.

The mine was found in 1657 by two brothers, Gaspar and José Salcedo. Ricardo Palma, a Peruvian essayist, tells in his *Tradiciones Peruanas* how José Salcedo had originally come to the hut of a native at the foot of the Laicacota hill, had asked for food and shelter, and had stayed and fallen in love with the daughter of the house, Carmen. When they married, Carmen brought the mine as dowry into the marriage.

It must have been a fabulously rich mine; it is said to have yielded 330 tonnes of pure silver in one year alone.[9] The Salcedos supposedly also founded a town and called it San Juan Bautista. The fame of the mine spread and attracted many people. Everybody wanted to get rich quickly. Another town grew, San Luis de Alva. Some 10 000 people are said to have lived there.

Quick wealth, greed, frontier spirit and lawlessness, immigrants of different origins who hated each other and fought each other all over the colonies during the seventeenth century – it all quickly led to tragedy. Rivalry between Basques and Andalusians caused continuous disturbances. The Salcedos were from Andalusia. They considered the Castillians and American-born creoles their friends. But Catalans, Gallegos and Vizcainos were enemies.

In about 1661, natives from the region of La Paz attacked and destroyed Laicacota, San Juan Bautista and other towns of the region. The troubles got worse and soon degenerated into a kind of civil war. In 1665, the governor of the province was killed. A Jesuit lost his life. There were continuous and bloody battles. In the absence of lead, bullets were made from silver.[10] Finally, in 1668, the new viceroy, Conde de Lemos, went up to the Lake Titicaca region to establish peace. José Salcedo was arrested, summarily tried and hanged. Apparently he was principally responsible for many of the disturbances. But perhaps it was his wealth which the viceroy coveted. Carmen disappeared into the mine; within three days it was flooded, the entrance covered with rocks – and no one ever found it again.

History has left us this bare outline of Salcedo's life. Was he guilty, and if so, of what? Could he indeed not buy himself out, being fabulously wealthy? And the mine, which had given so much silver, and had caused a settlement of 10 000 people – did it really just vanish overnight?

The viceroy had the town levelled and another one founded nearby. It was called San Carlos de Puno, in short Puno – on Lake Titicaca.

Baron von Nordenflycht

The baron was actually called Fuerchtegott ('fear of God', or Timothy) Leberecht ('live correctly') von Nordenflycht. His first names were strictly German, his last name Swedish. His father moved from Sweden to Germany in 1740 and then became adviser to the Polish King August III. Fuerchtegott himself went to the Mining School of Freiberg, graduated as a mining engineer and then became manager of a mine in Silesia.

His professional standing and reputation must have been extraordinary because emissaries of the Spanish king, Charles III, looked him up and asked for advice when he was still in his early thirties. As a result of this visit, Nordenflycht was hired as adviser to the Spanish crown and as travelling troubleshooter to reorganize the mines of Peru and Bolivia (then still known as 'Alto Peru'). It must have been a momentous decision for the crown.

For almost 300 years, practically from the moment when America was discovered, Spain's possessions had been off-limits for foreigners. Although quite a few non-Spaniards partook in the conquest itself, general xenophobia, religious intolerance, fear of intervention, of spies, and perhaps also an erroneous sense of protection, quickly brought about strict laws against foreigners. The growing importance of the Inquisition, and the persecution of Jews, converted Jews and Protestants – and even those simply suspected of heretic leanings – continued to keep America relatively free from immigration and even visitors.

Thus, the hiring of Nordenflycht was news. Change was in the air. Winds of unrest, of individual rights, and even of independence were blowing. Tupac Amaru, a native revolutionary and independence fighter had risen and had been suppressed only a few years before. Still, winds of change were one thing, but a German Protestant mining engineer going to Peru and Bolivia to inspect the industry, was quite another.

Fuerchtegott changed his first name to Timothy and, in 1788, left for Peru. He was accompanied by 13 German mining experts. Timothy was 36 years old and was to stay for 20 years.

They first sailed to Buenos Aires. The trip onward to Lima was one of extreme hardship. They travelled through the pampas of northern Argentina and the mountains of Bolivia and Peru, via Tarija, Potosí, La Paz, Cuzco and Huancavelica, to Lima – something like 4000 kilometres on horseback. After they had reached Lima, all kinds of things went wrong. Europe was way ahead of Peru. The liberalism of Europe was not accepted in South America, where the German Protestants of the Nordenflycht mission were still heretics. Denunciations were filed with the Inquisition.

Nordenflycht's mission to Peru would be called 'foreign aid' today, or 'foreign technical assistance'. And much of what is criticized of such missions today was already wrong then, about 200 years ago. The Germans were supposed to inspect, change and improve. They were supposed to know about cinnabar mining and amalgamation. And they were supposed to know how to organize. While they may have done some good here or there, history has not treated their memory very kindly. Their supposed excellent new amalgamation process turned out to be expensive (using more mercury than the one traditionally used in Peru) and impractical (using a lot of wood, which was, and remains, scarce in the mountains of Peru and Bolivia).

Comparative public tests were eventually made, under the supervision of the viceroy, and clearly showed that there was nothing the foreigners could teach the Peruvians. The traditional Peruvian process of amalgamation was cheaper and more productive.

A laboratory and the College of Metallurgy which Nordenflycht tried to set up swallowed about twice as much money as had been budgeted; the first was a failure, the second never opened.

Other suggestions and comments made by Nordenflycht and his colleagues showed a singular arrogance and lack of tact ('we, the foreign experts, will show you . . .'). Such attitudes did not go over too well with the Peruvians, and when Nordenflycht eventually left in 1812 the real results of the mission were considered few, and he had become a rather disliked expert. Or maybe there were other reasons?

Political matters probably caused Nordenflycht's failures; still, a spicy story always helps. He ran afoul of one Micaela Villegas, alias la Perricholi, a popular actress of Lima who had been the much acclaimed mistress of the previous viceroy. Nordenflycht's laboratory and 'metallurgical concentration plant' were near Micaela's house – up river. She and other residents sued Nordenflycht for using too much water. It took years before the suit was decided in his favour.

The nineteenth century

Between 1820 and 1826 Peru was convulsed by the independence movement. Peruvian creoles (Spaniards born in the Americas) fought side by side with Chileans, Argentinians, Colombians and Venezuelans for independence, against the remnants of the Spanish military forces. These were by then merely a faint shadow of the conquerors who, 290 years earlier, had defeated the Inca empire and established Spanish rule. Both sides, the independence fighters and the Spaniards with their loyalist followers, made generous use of large groups of conscripted natives, thereby reducing the labour force. The upheavals of those years brought most of what little mining activity there still was to a standstill.

This almost total stagnation did not end when Peru became independent, in 1824. A few years of rather luckless administration by Simon Bolívar were followed by years on end of self-proclaimed rulers, power-hungry generals, caudillos and plain adventurers. These turmoils did, of course, not encourage patient and productive mining operations. Off and on a few attempts were made, but without much success. It was also only about then, at the beginning of the second half of the nineteenth century, that really efficient pumps were invented. Until then, filtrations and accumulations of water in the lower levels of the mines were a constant problem. Often mines had to be abandoned because they could not be pumped out.

From about 1880, prospectors 'rediscovered' some mines in the central region of Peru, the most important being some old Inca workings called Cerro de Pasco. By 1887, these workings were considered a meaningful prospect. But before exploitation could really start, the railroad from Callao up into the mountains had to be finished. Work on the line had commenced in 1870, but had been completed only as far as Chilca before the government ran out of money. Then the War of the Pacific had interfered and it took until 1893 before the line reached La Oroya; it got to Cerro de Pasco only in 1904.

But exploration and surveying and sampling had been intensified when it became clear that the railroad would be completed in a few years and, in 1902, after securing large numbers of claims, a group of US investors decided to go ahead. They first formed the Cerro de Pasco Mining Company and then built a smelter at La Oroya, a rather narrow gorge where two streams and several roads converge and which seemed to be ideally located for quite a number of promising nearby ore bodies. Peru was, in terms of mining, entering the twentieth century with a bang.

La Oroya

Anyone who travels by train or car from Lima into the central mountains of Peru will pass, eventually, some 180 kilometres from Lima, through the little town of La Oroya. It is a sprawling, polluted and ugly place which owes its existence to Peru's principal smelting and refining complex. Many metals are produced here: copper, lead, zinc, bismuth, silver, gold, cadmium, antimony, selenium, tellurium, iridium, thallium, arsenic and tungsten. The La Oroya smelting complex is unique in South America and probably has few equals in the entire world.

Several aspects are outstanding. As the works are located at 3727 metres above sea level, the prevailing thin air caused, and causes, technical problems not elsewhere encountered. It is not only human beings who suffer shortage of breath and loss of strength at high altitudes. Many technical problems had to be overcome before efficient performance was assured. The many different metals which are produced at La Oroya point up another unique feature. The ores processed at the smelter are extremely complex and contain trace elements of most of the metals mentioned above. Many problems had to be resolved before all of these elements could be recovered. A further complication was that the ore feed comes from many mines and from all over the central mountain region. Where, by contrast, the Ilo smelter of Toquepala gets one type of concentrate from one huge open pit operation, or maybe two, the La Oroya smelter receives concentrates not only from the mines owned by the company but also feed bought from other companies. One of the most difficult jobs at La Oroya is to select and mix the daily feed for the furnaces.

The smelter began operations in 1922 and was gradually expanded. The first metal produced was blister copper with high contents of silver and gold. The copper was then shipped to the United States for refining and recovery of the precious metals. It was only in 1942 that Oroya got its own refining facilities.

The procedure with most other metals was rather similar. In the early years, different types of crude multi-metal bullion were produced and exported for separation and refining elsewhere. Company owned refining facilities were set up much later – antimony in 1941, arsenic and cadmium in 1943, iridium in 1947, lead and bismuth in 1949 and zinc and silver in 1952.

The La Oroya facilities belong now to Centromin, one of Peru's state owned companies. The owner has been the government, or the nation, of Peru since the beginning of 1974. This was when Peru expropriated, or nationalized, or brought about a forced sale by the previous owners, the US owned Cerro de Pasco Copper Corporation.

Cerro de Pasco

Generally, people referred to it as 'Cerro'. Back in 1902, it was 'Cerro de Pasco Mining Company'. Then it became, in 1915, 'Cerro de Pasco Corporation'. The word copper was added at a later date. It all came from the Cerro de Pasco mining complex which is, beyond doubt, one of the largest, most important and richest mineral formations in the world. It is located about 320 kilometres north-east of Lima, on a stretch of plateau between two chains of the Andes. At 4300 metres above sea level, the town of Cerro de Pasco and the mining complex there only exist because of the riches in the ground. No sane person would otherwise live at such an altitude and in such inhospitable surroundings. Close to 90 years of excavations and, in a way, 500 years of digging have not made the countryside more inviting.

Credit for the discovery of the mineralized zones – because this is not just one mine but many – must go to the natives long before the Spanish conquest. Then it was, and continued to be until the advent of this century, the lure of silver and gold. Old, pre-Spanish workings supposedly existed when, in 1630, outcroppings were discovered. No doubt the Incas had worked, and exhausted, the richest outcroppings and surface deposits. The means they had at their disposal to dig into the mountain were, of course, extremely primitive. Without adequate tools they could only 'break in' along natural and already loosened fissures. To tackle solid rock was beyond their capabilities.

The Spaniards were already better equipped. They knew how to apply fire, to break rock, and they had blasting powder. Also, their tools were better and, over the centuries, improved further. Workings abandoned by the Incas as 'worked out' or inaccessible became once again 'rich surface deposits'. The

Spaniards worked them for another 250 years. The confusion of the independence movement and the subsequent unrest of the early republican times affected the workings only temporarily, between 1820 and 1832. Thereafter the pursuit of silver from the Cerro de Pasco deposits continued with renewed vigour.

It must have been an unbelievable sight – hundreds of pits and mine shafts, each worked by only one or two men. They built their houses or shacks right there, beside, and even on top of, each mine. It made for good protection. And it shielded successful finds from the prying eyes of others. Much of the sprawling nature of what is, or was until a few years ago, the old Cerro de Pasco town, was caused by the haphazard building of shacks and demarcation of plots in the old colonial times. It is only since 1965 that major changes have been made. A new city has been built a few kilometres away, not only to give better living conditions and facilities to the miners but also, and this is the main reason, so that the open pit of the mine could be expanded to where the old city was.

During the War of the Pacific, 1879–83, Cerro de Pasco was occupied by Chilean troops. The mines were abandoned, in part because by then the grades had declined to a point where the people felt they had once more exhausted the rich 'surface' deposits, and the diggings into the rocks had reached limits of depth beyond which the mines could not go with the still primitive methods of the time. Water filtrations had already caused problems during colonial times, and the deeper the workings went, the worse it got. Repeated cave-ins and floodings claimed many victims.

Then in 1893 and 1897, the price of silver declined drastically, and so did the mining activities in Cerro de Pasco and, consequently, the fortunes of the city.

In those dismal days, towards the end of the century, several important events came together to give Cerro de Pasco and the surrounding mines not only a new lease of life, but finally to make it a mining centre of continental importance.

The first and foremost event to occur was that the railroad from Lima reached La Oroya in 1893. The line from La Oroya to Cerro de Pasco was completed only in 1904, but the lines from individual mines to Cerro de Pasco had already been built, beginning in 1869. Of equal importance was the progressive industrialization of Europe and the United States which during the second half of the nineteenth century caused increasing demand for copper. And copper was abundant at Cerro de Pasco. While the region had been considered the principal silver producer of South America since the decline of Potosí, the existence of substantial copper deposits was known – ore running 10–12% – but the deposits were ignored because the long haul by muleback or horse cart down to the coast made production uneconomical. But rail communication changed everything, as did the sudden involvement of foreign capital. This must be considered the second key event of those years.

The third reason which brought about the formation of the Cerro de Pasco Mining Company was a new Peruvian mining code which went into effect in 1901. It had been elaborated since 1896 by the newly founded Sociedad Nacional de Mineria and proved to be very liberal.

From 1887 on, a US syndicate had started evaluating the Cerro de Pasco copper reserves. An American mining promoter, James B Haggin, and a miner of Salt Lake City, Utah, A W McCune, acting on behalf of the syndicate, decided to send one James McFarlane, a mining engineer, on a fact-finding tour to Peru. McFarlane convinced himself that there was much copper and silver in the ground around Cerro de Pasco, and he convinced Haggin and McCune. They now set up, at the beginning of 1902, a New York-based holding called Cerro de Pasco Investment Company, which then acquired, through its Peruvian subsidiary the Cerro de Pasco Mining Company, most of the concessions of the Cerro mines, the Cerro–Oroya railroad, and the rights to other mines in central Peru. Among the shareholders was Michael Grace, a descendant of another Grace who had begun as a ship chandler in Callao.

William Russel Grace, the original W R Grace, was from Ireland and arrived in Peru in about 1850, poor and sick. A Dr Callagher nursed him back to health only to be asked by Grace for a loan of £100. He, Grace, wanted to set up a business and promised Callagher 50% of the shares. Callagher declined the participation and gave Grace an outright grant of £50, unaware that he was foregoing a half share in what would be, 100 later, a multi-billion dollar corporation.

Grace began ship chandling and then went into the guano business. The firm of W R Grace & Co was next set up, mainly to profit from the transport of guano; it eventually owned Grace Line. In its heyday, W R Grace & Co was to employ 15 000 people in Peru alone. The small ship chandlery became the cornerstone of an industrial and commercial empire which included Grace Line, 50% of an airline Panagra, two large sugar farms, a large paper mill, a chemical works, three textile mills, and a candy maker, as well as companies in the general export and import business, in ore buying, fishmeal production, paint making, mining, banking and industry in general. Grace's involvement in Peru lasted 120 years. Between 1960 and 1975 the company liquidated all of its Peruvian holdings.

William Russel Grace eventually returned to the United States and twice became mayor of the city of New York.

Michael Grace appears to have been a kind of godfather to Cerro de Pasco. He was apparently always the right man in the right place at the right time. In 1885, in the still confused aftermath of the War of the Pacific, his group picked up a controlling interest in the company that owned the still unfinished railways from Callao to Oroya and from Oroya to Cerro de Pasco. It also acquired ownership of the Rumiallana tunnel. This was the ambitious and long overdue project to drive a long drainage tunnel under most of the existing works at Cerro de Pasco. But the tunnel had yet to be built. All that existed was the project, and the right to build the tunnel. And that was what Grace acquired. But most probably he had no intention of completing the railroads nor building the tunnel. And, indeed, he ceded the railroad rights to one Thorndyke and exchanged his rights to the tunnel for shares in the Cerro de Pasco Mining Company.

But back to 1901. The shareholders of the investment company, which Haggin controlled by owning 34%, included J P Morgan and the widow of old George Hearst. In 1915, the Cerro de Pasco Mining Company and the Cerro de Pasco Railway Company were merged to form the Cerro de Pasco Copper Corporation.

In 1902, the first products of the new company were copper matte and silver bullion. The acquisition of many small claims spread over large areas was time consuming and difficult. The claims were often very small and much haggling, and investment, were required. McCune soon moved to Peru to take charge of the company's growing operation.

The Oroya–Cerro railroad was completed in 1904 and finally allowed the moving of larger quantities of ore to the two small and primitive smelting facilities they had at 'Smelter' and 'Casapalca', and to take the metal down to Callao for export. In 1905, Haggin acquired large claims in Morococha which, from about 1913, began to yield major quantities of copper.

Over the years, the Cerro de Pasco Copper Corporation, with the mines, the Oroya smelting complex, cattle farms, woodlands, railroads and many industrial participations, became an institution in the Peruvian scene. By the middle of this century it had become a sprawling giant which dominated the mining industry. By then it was serving as a training ground for generations of Peruvian mining engineers and geologists who went on to look for, find, set up, operate and expand other, mostly Peruvian owned mines. Cerro not only mined ores and produced metals; it also 'spawned off' a number of small and medium-sized industries 'downstream', those that used Cerro metals to produce alloys or semis, as well as those that produced tools and materials essential to Cerro's operations.

The sheer size of the operation made Cerro an ideal target for nationalistic attacks. Here was, after all, another foreign giant exploiting and exporting a most valuable national resource, metals, at what were obviously attractive terms to foreigners. And what was attractive to foreign investors had to be bad for Peru.

It would be tedious to review here the full story of how Cerro was slowly but inexorably strangled until the owners had practically no other choice but to sell out. The terms were negotiated throughout 1973 and on 1 January 1974 Cerro became the government owned 'Centromin'.[11]

How wise the expropriation or nationalization or 'acquisition by forced sale' was, is probably still debated in Peru today. It definitely contributed to the putting off of foreign investors for many years.

Nobody should doubt the sovereign right of a sovereign nation to take over or acquire any area or operation if it is considered of interest to the nation to do so. 'At a fair price' and 'with prompt payment' would appear to be self-understood conditions. No more sophisticated or ingenious argument would appear to be necessary. But that is really not what happened to Cerro in Peru. There was no fundamental reason why it had to become 'Peruvian'. The company obeyed Peruvian laws, paid all taxes, gave work to many, was a leader in housing and social assistance for its workers, and was technically innovative. Yes, Cerro had probably been slow, or negligent, in renewing equipment and blocking out new reserves throughout the years when the company came under ever greater pressure and ever more attacks. With the writing on the wall, who would have joyfully invested substantial additional sums?

But it was the overall situation and orientation of Peru, at first under the populist President Fernando Belaunde Terry, and then under General Juan Velasco Alvarado and his leftist fellow generals and civilian advisers, which doomed Cerro. As the 1960s progressed and became the early 1970s, Cerro's eventual 'Peruvianization' became inevitable – it was merely a question of when and in what form.

Let us hope that paying less than full value compensated the country for all the alleged and supposed wrongs Cerro had committed – as the government of Peru continuously proclaimed during 1973.

But our story has run ahead of itself.

The 1950 mining code

In 1950, a significant step in the development of the Peruvian mining industry came with the publication of a new mining code. It was in most respects a liberal, simple and workable code, establishing rules for private investment and opening the door to additional and substantial foreign participation. As a result, many new mines were put into operation, such as Toquepala (copper), Cobriza (copper), Marcona (iron), Millotingo (silver) and Milpo (lead/zinc). Exports of mining products rose in value from $45 million in 1950 to $454 million in 1968.

Toquepala and Cuajone

Toquepala must have been, for something like 30 years and until quite recently, one of the richest mines in Peru. It is a porphyric copper deposit in the southern sierra, inland from the port of Ilo. It is worked as an open pit. Its

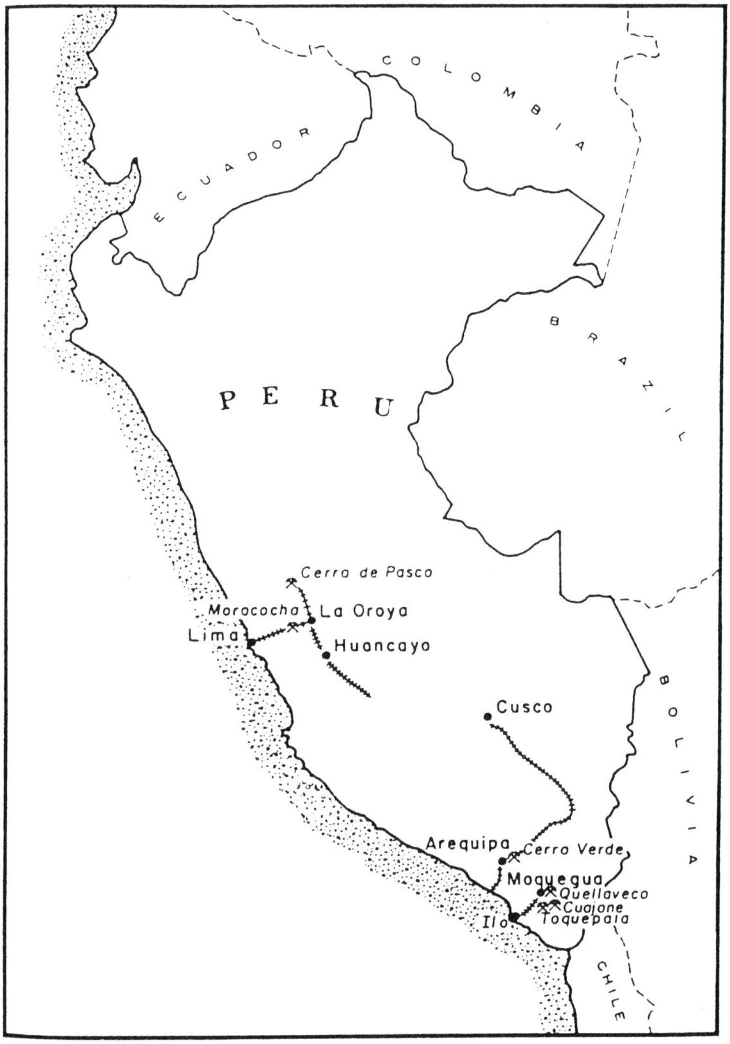

Railroads and principal mining centres in Peru.

existence was known for quite some years; a famous traveller of the nineteenth century, Raimondi, mentioned it. A Peruvian engineer registered the deposit sometime after World War II and offered it to the Cerro de Pasco Copper Corporation and to the Northern Peru Mining Company, owned by ASARCO. Both at first declined but eventually became interested in the property. A lawsuit resulted and both claimants ended up owning part of the new concession holder, Southern Peru Copper Corporation (SPCC). ASARCO held 51.5%, Cerro de Pasco Copper Corporation 22.5%, Phelps Dodge 16%, and Newmont Mining 10%.

It cost SPCC some $233 million to put Toquepala into production and that was, in 1960, big money. With the startup, Peru's production and export of copper quadrupled, from 51 000 to nearly 200 000 tonnes per year. The reserves were then calculated at about 400 million tonnes of ore with slightly above 1% of copper. Over the years, Toquepala has continued as a stable and steady producer of copper – about 120 000–150 000 tonnes of metal per year. In the mid-1970s, Minero-Peru, another government entity set up by the military government of General Velasco Alvarado that seized power in 1968, built a copper refinery at Ilo, and where SPCC had until then exported blister copper and concentrates, the refinery now converted the blister into cathodes.

When Cerro de Pasco joined ASARCO and the others to form SPCC, the company not only renounced its claim to the deposit in exchange for 22.5% of the shares, but also brought into SPCC its claim to Cuajone, a nearby copper ore body of rather similar nature and structure, even of similar size – 500 million tonnes of reserves containing 1% copper.

It took SPCC seven years, 1969 to 1976, and roughly $750 million to put Cuajone into production, again producing about 150 000 tonnes of copper metal per year and again as blister and concentrates.

Ownership of both Toquepala and Cuajone by the same group naturally facilitated much of the work necessary to open up Cuajone; a good part of the already existing infrastructure was used. Still, 242 million tonnes of overburden had to be removed before the open pit operation could begin. Twenty-seven kilometres of railroad tunnels were necessary. Fifty per cent of the total cost of the project went on new infrastructure.

It must be considered a miracle that the Cuajone project ever got off the ground and was completed. Planning began in 1968, the year the leftist military group around General Velasco Alvarado seized power. Work began in 1969. How come the generals did not expropriate Toquepala and Cuajone? The investment was so huge that even the generals with their otherwise limitless ambitions saw that the Cuajone project exceeded Peru's financial possibilities. SPCC then put together a financing package with loans from 50 banks in 10 countries on 5 continents. Expropriation would thus not just have hurt a single company but many banks in many countries. And, of course, the generals realized that expropriating Toquepala would have meant relinquishing all hopes of obtaining financing for Cuajone. So Toquepala was not touched, and Cuajone, the huge capitalistic and foreign venture, was started in Peru under a socialist regime.[12]

Not everything about the military regime of the early 1970s is bad. They must be given credit for putting Cerro Verde to work, for setting up the Ilo copper refinery, and for moving ahead with the zinc refinery in the Rimac valley near Lima.

The Cerro Verde copper mine

It must really have been at one time a 'green mountain', to be called Cerro Verde in Spanish. But the name probably did not come from lush vegetation, but from a greenish tinge of the earth caused by all the oxidized copper in the ground.

The mine is located close to Arequipa, a mere 4 kilometres south of the city, and at an elevation of 2700 metres. As copper mines go, it is neither a major ore body, nor a very rich one. The reserves totalled a mere 227 million tonnes of ore with 0.95% copper when the opening of the mine began, in 1973. These reserves compared unfavourably with Cuajone. But Cerro Verde was the first and, for a quite a while, the most important mining project of Minero-Peru, the state owned mining company.

The previous history of the deposit was, by itself, quite interesting. A subsidiary of Anaconda, the Andean Exploration Company, picked up the concession in 1916, and held it until 1970, when the 'revolutionary' government of General Velasco annulled the mining rights. The government first issued and then applied a new law: whoever held a concession had to put the mine into production, or lose the concession.

Anaconda had done some work between 1917 and 1918, and again between 1964 and 1967. Low metal prices between the end of World War I and the 1950s apparently made exploitation uninteresting. In the late 1950s, Anaconda wanted special conditions to open up the mine, and when the government refused to grant them, negotiations and plans for exploitation were again suspended.

Once the concession had reverted to the state, Minero-Peru picked it up and proceeded to put the deposit into production. By 1972 contracts were signed with Wright Engineers Ltd of Vancouver, BC and British Smelter Contractors Ltd to mine and concentrate the upper part of the ore body, calculated at the time to be about 29 million tonnes with 1.07% copper-containing sulphide ores. The estimated cost of this first stage was to come to about $55 million – all financed by Britain and Canada.

The idea was to produce copper, 99.97% pure, directly from the ore by a combination of vat leaching, solvents extraction and electrowinning, without smelting.[13]

It took Minero-Peru until late in 1977 and an expenditure of about $200 million, to get the mine into production. Inexperience, operational difficulties, but most of all technical problems, caused innumerable delays. When the startup time finally arrived, most outsiders seemed to agree that Minero-Peru would have been much better off ignoring the difficult to handle oxide ores on the surface and concentrating on the easier to process sulphide ore body which lies partially beside and partially below the oxides. By the time processing of the oxides began, the reserves of this type had been established at 61 million tonnes of ore with 1.01% copper, but a further 1.2 billion tonnes of sulphides containing 0.6% copper had also been found.[14]

Before taking a summary look at the period 1970–90, a few stories should be told about individual miners, mining families and specific companies.

Alessio Sibille

Alessio Sibille died in about 1980, at age 84 or so. By then he was living in Lima, as his days of work and life in the sierra were over. A few years earlier, the altitude had begun to bother him and it was then that he had moved to Lima for good. Still, he and his life had been 'sierra'. The mountains had made him; he had made it there.[15]

He was born in Turin, Italy, in 1896. In 1914 that made him just the right age – 18 – to have to join the Italian army. Five years of service followed. He made it to lieutenant. Then what? Those who left the service received at the end of the war double pay for the five years, plus one year's salary as a bonus.

In 1918, the number of those looking for a job was enormous, and jobs were few. For a while, Alessio studied agronomy and a bit of veterinary medicine. But conditions did not improve. Would it be better elsewhere? The US Consulate had a long waiting list. A distant cousin lived in Callao, Peru, and worked there as a teacher at the local Italian school. Upon enquiry, the cousin wrote that there was 'lots of work'.

The immigration formalities were easy; the trip took 35 days from Genoa to Callao. Sibille hid his savings in a small cloth bag under his shirt.

The cousin knew all the members of the Italian colony of Callao, and he introduced the new immigrant. The Lercari and Piaggio families were prominent first or second generation Italo-Peruvians of Callao. Ernesto Lercari owned two sheep farms in Cerro de Pasco and had use for a good agronomist with veterinary experience. So Alessio, after only a few days in Callao, took the train up into the sierra and worked as the administrator of a farm, from 1921 to 1925.

To work for others was not his life's ambition. He wanted to be his own boss. It appeared that there was good money to be made, even if one had to begin on a small scale. In 1925 Alessio opened, in the city of Cerro de Pasco, a small general store and a *casa de prestamos*, a money-lending house. The business developed well, between 1925 and 1929, and he bought out the owners of one of the largest general stores in town, Casa Azalia, on the main square of Cerro de Pasco. He dealt in all kinds of merchandise and also acted as commission agent. With hired help, and eventually with one of his sons in charge, Casa Azalia continued in the family until 1970, when they finally sold out.

Buying, selling and bartering goods from a general store in Cerro de Pasco quickly brought substantial profits. Never a spender, always frugal in his lifestyle, Sibille was soon faced with the question of what to do with his profits. He decided to buy a small farm on the outskirts of Lima. It became quite valuable as the city grew and agricultural land was urbanized. He also bought a farm near Huanuco. It was eventually expropriated, at the time of the land reform of General Velasco, in about 1970.

But Sibille was also tempted to go into mining, especially as mining picked up again in the late 1930s and people once more made money in it. He first went into coal from 1946 to 1960. But it did not work out. The grade was poor and the mine too far away. It was not economical to transport coal on muleback across the Andes.

When the Peruvian government passed the new mining code in 1950, the prospects for profitable ventures improved substantially. Better metal prices due to the Korean War also helped. Late in 1950, with two partners, Sibille staked out a lead/zinc/silver mine. A year later he went to the United States to buy a small concentration plant which they set up at the hamlet of Quilacocha; they called the operation Concentradora Quilacocha SA. It worked from 1951 to 1975. The partners made money with the mine, but they did not have much of a bonanza, so nobody got rich. The workings were never more than a hobby. By 1974, the partners decided to give up. Some of the claims were sold to Cie des Mines de Huaron, others to San Gregory. And in 1976, they sold the plant to Minera Cerro de Pasco.

The Baertls and Milpo

Ernesto Baertl Schutz (1892–1970) was, between 1950 and 1970, a big and important man in the Peruvian mining industry.[16]

Whoever met him once never forgot him. His was a powerful personality. He was rather short, but solidly built. In his younger years, his hair was supposedly flaming red. His features were kind of angular. He appeared a bit on the gruff side – especially if one visited him as an ore buyer. But he was a kind man, behind what was usually an inscrutable face. One occasionally got the impression that his appearance of toughness was just put on, on purpose. If he knew someone was German, or knew German, he would occasionally put in a phrase or two in that language to show that he had not lost all contact with his background. But, of course, he was Peruvian from birth to death, and among the best men Peru has produced.

He was, actually, a first generation Peruvian. His father – the original Baertl (1860–1917) – had come from Ulm in Germany. One of the old man's early friends from Ulm was Alex Oechsle, he who set up the Oechsle department store in Lima. Old man Baertl was a saddlemaker. Nobody knew why he had come to Lima, nor when. For years he had his shop on Jirón de la Unión, next to the La Mercéd church.

Ernesto studied at San Marcos University, then at the Escuela de Ingenieros. After graduation, in 1914, he worked for the next two years with the Cuerpo de Ingenieros de Minas on a survey that required travelling all over the mountain regions of southern Peru and Bolivia.

He spent the time between 1915 and 1921 in Castrovirreyna – one of those forbidding, thin-air, high-up regions of central Peru – at a silver mine called Santa Inez. To get to Castrovirreyna in those days was quite a task. Though the town is only about 200 kilometres from Lima the way a plane flies – it took four days. The first would be spent going by train to Huancayo, the second on horseback to Izcuchaca, the third to reach Huancavelica, and finally the fourth day to reach the mine. The travellers would sleep under the most primitive conditions in huts along the way.

Some of the peculiarities of day-to-day life in former times should be recalled, such as what transportation and communications in general were like then, to better understand those days. When travelling today, we take roads, bridges and tunnels very much for granted. We quickly grumble about a poorly kept road, a blind curve or too steep a climb. Rarely do we ask ourselves what people did before the road was there; how people got from place to place. Previous generations were generally used to riding on horseback, three days on a horse cannot have been much fun for Ernesto Baertl – nor for anybody else – in those regions.

One of Baertl's jobs in those early days in Castrovirreyna and before implantation of selective flotation in the early 1920s, was to travel to some salt flats near Ayacucho to organize shipments. Considerable quantities of salt were required as a chlorinating agent in what was called the 'toasting process' of silver bearing ores, to get silver chloride, to facilitate precipitation and amalgamation. The amalgam was then heated and much of the mercury recovered; the resulting product was silver sponge.

The Santa Inez mine produced what were then called, in Spanish, *quesos de plata*, or *plata piña* ('cheeses of silver' or 'pineapple silver', names doubtless derived from the shapes of the pieces). Mules took these 'cheeses' down to Ica; hence they went by train to Pisco, then by boat to Callao, and on to the mint in Lima. Or they went for export by steamer to Europe or the United States.

In 1921, Eulogio Fernandini Clotet (see pages 106–107) – a classmate – hired Baertl as a metallurgist for Huaraucaca, but only a year later the smelter there was closed. For reasons no longer clear, it became cheaper to send the concentrates to Cerro de Pasco's new smelter at La Oroya rather than to continue to operate the smelter. Baertl next worked at the Colquijirca mine (also owned by Fernandini) which supplied the Huaraucaca smelter. In 1925 he became the administrator.

The 1929–31 depression then also reached Peru and eventually affected Baertl; operations at the Colquijirca mine were shut down and the staff laid off. But Eulogio Fernandini kept Baertl on as watchman with, however, a cut in his salary to half the pre-depression level. Conditions only got back to normal in about 1934–5.

Baertl stayed with Fernandini until 1942, when he resigned and moved to Lima. He was then close to 50 years of age and had spent a lifetime in the mountains. But his health was excellent, good for another 28 years of life and for another career.

One wonders what made him give up his job choose a new life at this stage. Maybe he had had enough of mining camps and the primitiveness of life in the mountains? Maybe it was concern for his wife, that she should have a better life and the children good schooling without splitting up the family? Or maybe it was a longing for the life in the capital of the country. Or maybe it was a combination of all these things.

There he was, 50 years old, with a wife and six children – and no job. It apparently did not bother him. He knew what he knew, what he was worth, and that he was respected and liked. It was thus probably not too daring a step after all.

Felipe Ayulo, Felipe Beltrán and Leon Rosenshine – three entrepreneurs of significance in the Peruvian mining scene of the period 1940–70 – owned Volcan Mines and Castrovirreyna Metal Mines. Upon hearing that Baertl was available, they at once asked him to work for them. He accepted and was made general manager. And travelling into the mountains was also part of the job.

A few years earlier, Augustín Arias Carracedo had discovered metallic outcroppings near where Cia Minera Atacocha had begun operations some years earlier. Some ore was handpicked. Ernesto Baertl and several friends now took a first, then a second, and finally a third look, spent some money to explore further, and then decided to secure the claims. In 1949 they formed Cia Minera Milpo SA, pooled their funds, and began to work. The group of friends were Ernesto Baertl, Aquiles Venegas, Axel Nicander and Luis Caceres. All four had, at one time or another, worked for Eulogio Fernandini. They also took in Manuel Montori, a relative of the Baertls. At first, they alone owned all the company, but when additional funds were required, they invited Eulogio Fernandini to join, and he did. Luis Picasso, Domingo Bezzola and Augustin Aliaga also joined.

During the last 40 years, Milpo has acquired a reputation as a modern, well-run mining company. Ernesto Baertl eventually turned the presidency over to Luis Picasso, while Aquiles Venegas turned the management over to Baertl's eldest son, Ernesto (Tito) Baertl Montori. In time, Venegas, Baertl and then also Picasso died. Tito Baertl remained in charge, and in time turned the company over to the next generation. The shareholding by now is probably such that at least 51% is controlled by the Baertls, the Picassos and the Montoris. The Bezzolas and the Fernandinis probably also still hold about 5% each; some 15% is in the hands of workers, executives and staff, and the rest in the hands of the general public.

Milpo continues with its reputation for excellence, in planning, management and labour relations. Old man Ernesto Baertl Schutz is remembered as a simple, competent man with a big heart. He is also remembered for his uncommon willingness to help others, and to give free advice. Economic gain, personal enrichment, and a fee for advice were of secondary importance to him. Thus, though he was a big man among the many big men in the mining scene of Peru during the years 1900–70, he was one of the few who did not amass a big fortune. Though no doubt well off at the time of his death, and a good provider for his wife and children, he apparently left no fortune to them. But he did leave his imprint on a number of companies: Milpo, Pacococha, the Fernandini group, Volcan, Castrovirreyna Metal Mines and Cia Minera Castrovirreyna (Caudalosa), Chapi, and the organizations of Mauricio Hochschild y Cia Ltda and W R Grace & Co.

Three of his sons continued in the business. Tito, probably more than Augusto and José Antonio, did step fully into old Don Ernesto's shoes; in fact, Tito probably surpassed his father in achievements. Tito eventually served as president of Banco Minero and the Peruvian Mining Society, as well as on the boards of some 10 or 12 companies – as had his father before him.

Zacarias and the Millotingo mine

In 1979 one could still find a few old friends who greeted Felipe Zacarias Alvarez as 'Felipito'. He had been called that in his younger years, to differentiate him from his father, Felipe Zacarias Montañez. Felipito means 'little Philip'. But there was not much about him that was little. Though not very tall, he was not really short; he stood taller than his father. In business acumen, as a professional mining engineer, and as a successful executive, he had few peers in the last 50 years.[17]

Felipito looks about 60 years old, maybe 65. Actually, he is much older, for he was born in 1907. He still looks like a tough man – every inch a miner of the Peruvian sierra. He blends easily into the racial mix and outward appearance of the Limeños of today.

Felipito is Peru. The best of it. Self-made, both he and his family. Hard-working. Much of his life has been spent in the mountains and the little sierra towns. He has been steeled by the coldness and dryness of the mountains, by the sierra air, the dusty roads, the tunnels and the galleries of the mines, and made by flotation cells, crushers, separators, by dealing with ignorant labourers, agitators, and with the poor peasants of the mountains. He has also been moulded by the fight of generation after generation against the peculiar nature of the Peruvian mining regions: to scratch, hammer, chisel, dig and dynamite a living, a better living, out of the rocks.

Cia Minera Millotingo SA *was* the Zacarias. They made it; it made them. The Zacarias own other mines as well, also real estate in Lima, quite a bit of it, and they own, wholly or in part, other businesses. But nothing is, or was, quite like Millotingo, or Millotingo's two mines, San Juan and Siete de Diciembre. It is a story of silver, tough luck, good luck and wealth.

Felipe (Felipito) Zacarias Alvarez is today *the* Zacarias. It is just that this be so. It was, after all, he who really opened up the silver mine, operated it, and made the major part of the family fortune. But it was the father, old man Felipe Zacarias Montañez, who laid the groundwork. Though he is gone, the story should begin with him.

The old man died in 1981, at age 96 or so. Nobody knew his birth year. Maybe nobody ever knew. Perhaps there still are, somewhere in an almost forgotten town in the sierra, barely legible records showing when he was really born. But it would hardly matter. His was a full life, and precise dates were not important. Precision in mining and finances were what mattered. And in that he excelled.

Old Don Felipe was born some time around 1883. He went to school in San Ramón, near Tarma, and at one point won some kind of gold medal for 'excellence'. Eventually, the family moved to Yauli. He worked with his father – whom we shall refer to as grandfather – who had a general store, selling everything from food to tools. In about 1912 they moved to Morococha, and opened up another store.

There was not much in those regions other than sheep-herding and mining. And the miners – owners and workmen – depended on those small stores for most of their supplies. Grandfather Zacarias and his son Felipe – Felipito has just commenced school – met and traded with the people of the nearby Sacracancha mining company. With time, closer relations developed. The Zacarias supplied the mine, and in turn represented the mine in Morococha. As their interest in mining grew, they picked up some mining claims. In around 1916–17, they – grandfather and Felipe – staked a mine called Fortuna and began working it on a small scale. It gave high grade silver ore. The ore was sold to what was then called the South American Metal Company, a subsidiary of the American Metal Company, which was buying supplies in Peru as agent for ASARCO, the American Smelting & Refining Company.

In around 1925, grandfather was sued by a lady he had lived with for many years, and by her relatives. It must have been a spicy story, were one able to piece it together again, but it happened too long ago, memories are dimmed and papers lost. Grandfather had apparently lived with the lady for years but had then decided to call the arrangement off. The affair supposedly never had been formalized – so the story goes – but she claimed they had been married, that grandfather had abandoned her, owed her support, and, as grandfather was now making money, he should share. She sued grandfather for a part of the ownership of the Fortuna mine.

Law, law enforcement and interpretation were iffy matters in those days, and grandfather's case was shaky. What hurt him most was that the page of the parochial marriage register in which the lady claimed the event had been inscribed, was missing. Somebody had torn it out. Whodunit? Grandfather, to destroy proof of the event, or the lady, to be able to rely on the sworn testimony of false witnesses? Grandfather lost the suit. But as all of the legal back-and-forth had left no time to work the mine, the claim lapsed and an outsider, Dallago, took advantage of the confusion, filed a new claim, and worked it for the next five years. The Zacarias were again out of the mining business, but the real loser was the lady. She had won the lawsuit, and was supposed to share in the proceeds from the mine that the Zacarias worked, but could not, of course, claim anything from Dallago.

For a while the Zacarias then worked another mine called Zoraida. They still had the general store, and continued to search for new mines.

In 1925, Felipe went to Lima and bought a service station which, for several years, was a moderate money maker before he sold it. He also set up a company which ran buses as part of the Lima transportation system. But life in the big city was not for him. He could not leave mining alone. In the late 1920s, he decided to work a mine called Viejecita, in the Carahuacra district. It belonged to Fermín Málaga Santolalla. They went 50/50 on the profits, with Zacarias doing the actual mining. On his own, Felipe put up a concentration plant in Yauli to treat the ore. But when, after a while, profits became sizeable, Málaga decided he might as well work the mine alone, without a partner and without sharing the profits. And as no written contract had been drawn up, Zacarias was out as partner of the mine and his plant had no feed.

Well, maybe the story was that way or maybe it was somewhat different. But if it is not told here the way it was, it could have been this way. Who can say today what influence the economic crisis of the early 1930s had on these matters?

In 1933, Felipe went to work as a mine captain for Negociación Lizandro A Proaño. Felipito, then aged 26, who had by now graduated from engineering school, went to work for the same company. Felipito's first job was to put up the Tamboraque concentration plant.

The two Zacarias only stayed with Proaño for about two years. Every Sunday, to relax and to get away from the camp, the two roamed the surrounding hills looking for mines. They were driven by the dream of every mining man that he is always about to find the one big and profitable mine to which every miner believes himself entitled at least once in a lifetime.

Felipe and Felipito Zacarias looked again and again. And eventually, in 1933, they found the San Juan mine, which years later became Millotingo. It had previously been considered a good prospect. A man by the name of Moscoso had owned the claim; he had tried but had failed. His claim lapsed. The Zacarias registered a new claim. They opened a hole, dug an adit: ore had to be there, and in quantities, and of good grade. They advanced some 20 metres, with pick, shovel and some dynamite. But the grade remained poor, only 8–10 ounces of silver per tonne. They gave up. Most of their savings were gone. It was back to work for others. Felipito hired himself out to Angélica de Osma as engineer.

Angélica de Osma's story itself is quite remarkable. It goes back to a man called Pedro de Osma, who was born in Lima in about 1867. He studied law, became a deputy for the department of Cerro de Pasco, later for the department of Lima, and still later president of the Chamber of Deputies. In 1903, he founded the newspaper *La Prensa*. One day, mining caught his fancy, and in 1926 he set up Sociedad Minera Yauli SA to exploit some lead/zinc/silver mines in Morococha. From his marriage to Angélica Gildemeister two children were born, Pedro de Osma, known in the 1950s and 1960s for his priceless collection of Peruvian antiques and memorabilia, and Angélica de Osma, who dedicated herself to running Yauli, which still today remains an important producer of zinc and lead concentrates.

Felipito worked for Yauli, starting in 1934, while his father Felipe was looking around for something to do. He still had the concentration plant he had built for the ore from Málaga's Viejecita mine. Málaga now wanted to buy it and offered Felipe 5000 soles. But Felipe wanted 10 000 soles; they could not come to terms. However, Felipe had another idea: he would give Málaga the plant in exchange for the mine. They agreed. It is unclear what Málaga did with the plant. It was probably taken elsewhere to process ores of other mines.

Felipe now had Viejecita and worked it on a small scale. The old Fortuna mine also became free again. The Zacarias filed a new claim and began working it. The old lady who had pursued grandfather had held on to her judicial award and was given 25% of the shares. The Zacarias had the remaining 75%. But Fortuna's fortune was gone. The veins petered out and they soon abandoned the workings. The old lady was out of luck one more time, and this time for good.

Some of the folklore of the Peruvian mining industry has it that the Fortuna mine made the fortune of the Zacarias. But this was not true. By the late 1930s, when the Zacarias abandoned it, they were still poor, if not exactly destitute. Dallago, who had worked the mine before them, later founded the Sindicato Minero Rio Pallanga.

The next significant step came in 1944, when Banco Minero bought the Sacracancha concentration plant from the Bozas. The independent miners of the region now had a custom concentrator nearby. And that included Zacarias and the ore from Viejecita. Felipe Zacarias could step up the production of selected silver ore and work the mine on a somewhat larger scale. About every five months, rather substantial accumulations could be processed in a 'concentration campaign', resulting in concentrates that went to the Oroya smelter of the Cerro de Pasco Copper Corporation. Each of these campaigns left a profit of about 50 000 soles, or about $8000 – a lot of money in those days.

The Cerro de Pasco Copper Corporation now became interested in Zacarias' operation. It was then in the process of buying up claims in the Carahuacra region. A geologist of Cerro de Pasco looked Felipe up and offered him 50 000 soles. After all, he thought, a meek-looking miner should not have large monetary pretensions. But the geologist was in for a surprise. Felipe told him very plainly that there was a mistake in the monetary sign; it should be dollars instead of soles. Cerro de Pasco bowed out, but Ernesto Baertl and Cy Fleischmann became interested, on behalf of Volcan Mines. They paid the $50 000 for Viejecita, $20 000 in cash and the balance in three instalments of $10 000 each, every six months.

Thus, in around 1945, Felipe had $50 000. A large stake: what to do with it? He then remembered the San Juan mine. He still had hopes, as had Felipito. Now they had money. The claim was free and they registered it in their name. Work began. They drove further into the hill, tenaciously, with patience and hard work. The grade of ore improved. At 60 metres, a grade of 60 ounces per tonne was obtained. They began to ship about 30–60 tonnes a month to the Oroya smelter. Work continued at about the same rhythm between 1946 and 1952. They made money, but no real wealth.

In 1952, Felipe's wife died. He was now 62 years old. Felipito thought his father had done enough hard work and suggested that the old man should take life easier. He should stay in Lima, not go up to the mine any more. He should leave the work to underlings, and to Felipito's supervision. But the old man had different ideas. He flew into a towering rage and accused Felipito of considering him old and useless – and for the next two years father and son did not speak to each other. The old man continued to operate the mine, San Juan, and Felipito ran Yauli. In 1954, the old man finally gave in, called Felipito and asked him to take charge of the operation. Felipito did. The operation had just been coasting along, and continued that way for another two years.

But luck was again with them. In 1956, the road from San Mateo to the Tonsuyoc concentration plant was finished and the mine was literally right 'on its doorstep'. Overnight, transportation costs became cheaper and larger volumes were possible. They also began to ship to Oroya handpicked ore containing 80–90 ounces of silver per tonne. Systematic exploration gave them by 1957 about 45 000 tonnes proven reserves, with an average of about 30 ounces per tonne. They decided to build their own concentration plant, with a capacity of 100 tonnes of ore per day. It was a major expense. People said they were crazy. The reserves were barely enough to work the mine for a year. One does not take such a gamble. But they were stubborn and confident. They went ahead. The plant began operations in 1959. One year later they had put through 10 000 tonnes and, in addition, had increased the reserves to 90 000 tonnes.

Until the end of 1959, Millotingo produced some 600 000 ounces of silver (about 18 tonnes), contained in concentrates. In 1960, capacity was increased to 250 tonnes of ore per day. Production now jumped to an average of about 1.5 million ounces per year (about 45 tonnes). Their best year was 1962, with 2.3 million ounces of silver (about 70 tonnes) and 8000 ounces of gold (about 240 kilograms).

However, the heads (silver content of the ore taken out of the mine) gradually decreased to the 13–14 ounces per tonne range and 1975 was a bad year, the first in which Millotingo lost money. To the already low heads, and bad economics, were added two lengthy strikes, political turmoil because of the government-imposed Comunidad Minera, and generally higher costs due to spiralling general inflation.

By 1991 production was still continuing, at about 25 tonnes of silver a year. But Millotingo is no longer profitable. The lush days of the early 1960s are over – until, perhaps, new, rich veins are found. Felipe has long since passed away, but Felipito still runs the company, and the group, as other mining properties have been added.

The Pflueckers and Morococha

The Pflueckers were on the Peruvian mining scene between 1820 and 1920. They burst upon it quite suddenly, became probably the most important pioneers of Peruvian mining in the early days of the republic, became extremely wealthy, and then faded away.

Throughout the centuries, the boundaries in south-eastern Europe have suffered many alterations. And the racial and national origins of the inhabitants have often not kept up with political changes. Silesia was Prussian and German from 1742 to 1945. But not all of its inhabitants were German. The Pflueckers of Breslau, the largest city of Silesia, were Polish according to the sources. But their family name as well as their first names are so thoroughly German that they probably were that – German. Otherwise, why would a Pole have given his children German names?

In 1820, Carl and Julius Pfluecker left Breslau to go to Peru. They wanted to work some of the old Spanish silver mines in the mountains there. South America was throwing off the yoke of Spanish colonialism; now was the time to 'get in on the ground floor'.

Carl was a mining engineer; Julius had studied commerce and economics. Their father, a prosperous 'Polish' banker in Breslau, had seen to it that they got a good education.[18]

The Pflueckers began by searching in the Huancavelica region. Their first workings were at a mine called Quizpiziza, near Castrovirreyna. They got the mine cheaply and quickly turned it into a money-making operation. But the real money-maker was the Morococha mine, which they bought in the late 1840s and from which massive production of rich silver ore began in the late 1850s. Meanwhile, both brothers acquired families.

The full story of the rise and decline of the Pflueckers has never been told. It would be a fascinating tale. From the fragmentary information available, one gathers that their real bonanza lasted at the most 15 years, between about 1860 and 1875. By the late 1860s, the Pflueckers were known as the 'millionaires'. They had acquired houses in England and on the continent – besides houses and land in Lima – and several of their children were educated in Europe. In about 1871, Julio, one of Carl's sons, married Hortense, one of Julius' daughters. Julio in time took over the management of the mine and continued to reap further riches. But it all ended with the next generation, under Roberto, son of Julio, who had been chosen – apparently more by default than special inclination – to run the operation. He hardly ever visited the mines. He and the other descendants preferred to spend the income, and then the capital. Administering it all was just too much trouble.

In 1915, when the Cerro de Pasco Copper Corporation bid 2 million soles for the Morococha mine, Roberto sold out.

The family still owned much real estate in Lima, sugar farms and other investments, but the real money-maker had been the mine, and it was now gone. Admittedly, the rich silver veins had long ago petered out, but the mine still produced a profit and was again to produce a fortune, though in copper, for the new owners.

The pioneering spirit of the first Pflueckers did not survive even into the third generation. When, in the early 1920s, Roberto Pfluecker died, the last of the male Pflueckers disappeared. By then, most of the money was also gone. Nothing much of value was left for his only daughter. And Roberto's last days were spent in the Magdalena asylum. He had become a hopeless dipsomaniac.

We followed the fortunes of Alessio Sibille, an Italian immigrant who bought a general store on the main square of Cerro de Pasco and, as well as lending money, also dabbled in mining, although not with too much success. The owners of another store in the city of Cerro de Pasco were more successful. The following is their story.

The Gallos and Atacocha

'Gallo Hermanos' belonged to immigrants from Spain, the Gallos and the Diez Gallos. Theirs was also a general store on the main plaza. They also dealt in all kinds of merchandise, anything that was needed and could be sold at a profit. In 1925, one of them, Gerardo Diez Gallo, filed several mining claims. They were generally referred to as Atacocha, a mine located but a few kilometres from the city of Cerro de Pasco. The family parleyed those claims over the next 60 years into one of the best-known and respected medium-sized mining companies of Peru.

It appears that the first reference to the ore deposit of Atacocha was published in 1828 by Rivero y Ustariz in his 'Memoria sobre el Rico Mineral de Pasco' (Report on the rich ore of Pasco) in which the author mentions a hill called Chuquitambo with some gold mines, and several other mines called Quinua, Chiquirin and Huamautanca. Chiquirin must have been Atacocha. Antonio Raimondi, the most interested and dedicated of the nineteenth century foreigners to travel in Peru, added in 1867 more detailed comments, including descriptions of ore samples he took and some assays. But, although the presence of worthwhile silver contents was determined, nothing happened at the ore body for another 70 years.

'Nothing' is not quite correct. The region was surveyed and mapped. Luis F Diaz, a mining engineer who worked for the Mining Engineers' Corps to produce a first and basic register of mining claims, pointed out Atacocha as of considerable potential. By 1908, what was later on to form Atacocha was divided into 29 claims. The Three Musketeers mine and the Dora mine were already worked; however, on a small scale.

In 1907, some 900 tonnes of ore were extracted by several small mining operations. The most successful of these belonged to J H Fleming, a former superintendent of the Cerro de Pasco Mining Company. In 1915, Fleming joined forces with other ex-Cerro officials, H Rally and J D Torbert, to form a company which they called Pucayacu Mining Company. Some of the operations were mechanized – by means of jigs. They produced a high grade lead ore, as well as silver and gold. One of the curious details about the mining activity in the Atacocha area was an element of speculation. It was known that the Cerro de Pasco Mining Company – the big US-based mining group – was buying up claims for a major consolidation. The miners of the Atacocha ore body speculated that their claims and mining would be bought up – surely at an overprice – for incorporation into the big company. But when the Cerro de Pasco Copper Corporation was formed, Atacocha was not included. Many of the small operators then lost heart and gave up and the ore body entered once more a period of neglect. Soon thereafter Fleming died and the Pucayacu Mining Company went into liquidation.

When, in 1925, the claims became free, the Gallos filed them anew in their name. They invested substantial sums for exploration and tunnelling and in 1928 managed to get the Cerro de Pasco Copper Corporation to take a closer look. Also Mauricio Hochschild y Cia Ltda, a Peruvian ore-buying house owned by one of the tin kings of Bolivia, studied the ore body in 1929. But the depression intervened, demand for metals dropped, as did prices, and yet another protracted delay occurred.

It was finally in 1936 that a new – and finally successful – attempt was made. The Gallos formed Cia Minera Atacocha SA. In time it became one of the largest privately owned mines of Peru.[19]

Its production, in 1990, reached 13 252 tonnes of lead and 22 465 tonnes of zinc (metal content contained in concentrates) as well as 1.5 million ounces of silver (about 45 tonnes).

The Fernandinis and their mines

The name Fernandini is intimately connected with the Peruvian mining scene of the twentieth century. The man responsible for this was 'old Don Eulogio' – Eulogio Fernandini de la Quintana – who, between 1900 and his death in 1947, built up another medium-sized and fully Peruvian owned mining group.[20]

The origins of the Fernandinis are lost in history but, judging by the name, they probably came from Italy. However, nothing is known of who was the first to come to Peru, and when, and why. Apparently an early member of the family, a general called Juan Pablo Fernandini, was executed in Arequipa on 18 February 1836, together with General Felipe Santiago Salaverry and others, in a reversal of one of the many civil wars which followed Peru's independence.

Eulogio Fernandini de la Quintana was born in 1860 in Ica, the son of a lawyer from Lima. He spent some time in Germany, apparently to complete high school studies, but he was back in Peru in 1880, by which time his father was a judge at the district court in Cerro de Pasco. Eulogio's first job was with the Gallo brothers, general merchants and, in a way, merchant bankers in Cerro de Pasco. In 1884, he married Isolina Clotet Valdizán, an 18 year old girl from Lima. Part of her dowry was in cash and this apparently allowed Eulogio to begin work as an independent miner.

These were the years when, among others, quite a few Americans became interested in the mines around the town of Cerro de Pasco, and the constant coming and going of engineers, the filing and refiling of claims and the planning for a major mining complex excited the people of the region. Fernandini also became interested.

For a number of years he spent much of his time dabbling with small mining claims, moving about, with his wife, in the towns and camps of the Cerro de Pasco region. It may have been due to this harsh life that the first five children of the couple died in infancy.

Among the many abandoned 'holes in the ground' – leftovers of old Spanish workings – were those of a mountain named Colquijirca (colqui =

silver, jirca = mountain). Fernandini and a few partners began to work these holes with drifts and adits, until they struck rich veins.

In those days (1902–3) only silver ore made commercial sense. And it had to be high grade ore, because transport was primitive and expensive. The ore had to be bagged and taken on mules or llamas over difficult terrain.

Colquijirca turned out to be a rich and profitable mine. After a few years, Eulogio Fernandini could afford to build his own smelter (instead of sending the ore to La Oroya). It was named Huaraucaca. The resulting lower transport costs increased profitability, allowed the processing of lower grade ores, and permitted larger quantities. In no time, Colquijirca became one of the principal producers of silver in Peru. Fernandini and his company even received permission to mint coins.

The type of ore found in those early years at the Colquijirca mine was very unusual. Anyone who saw it always remembered it. A few samples were kept for decades but were eventually lost. The ore looked like white hair – innumerable, interwoven strands of pure silver. The ore was called *cabello de mujer* – the hair of a woman. A tonne of it often contained upward of 200 ounces of silver.

The rich veins of the Colquijirca mine eventually petered out; the ore contained less and less silver. The mineralization then changed to a much more complex ore, containing lead, silver, gold, zinc, copper and cadmium. The greater complexity of the ore required ever more sophisticated methods of concentration and separation.

And still by the early 1990s, the search for optimum recovery methods continues. Operational control of the company, El Brocal, has meanwhile passed to outsiders, managed by Alberto Benavides (formerly of Cerro de Pasco, and also co-owner of his own sprawling mining group, Cia Minera Buenaventura SA, the second largest silver producer of Peru).

Surprises are always possible. The ore continues to be extremely complex and very difficult to separate. What else will the future bring? In 1960, it should be said, they suddenly found a *clavo de plata* (a silver nail) of some 25 000 tonnes of ore with an average of 120 ounces of silver per tonne.

Eulogio Fernandini worked the mine successfully and on an ever larger scale, from 1902 to 1931. We should not doubt that he obtained very handsome profits over those years and accumulated a large fortune. Much of this fortune was reinvested, in other mines and ventures. For example, the smelter needed coal. So they looked for coal, found a prospect, had the area staked, and Fernandini had a coal mine.

In about 1920, old Don Eulogio's eldest son, Eulogio Fernandini Clotet, introduced selective flotation. Now they could expand further, as lead and zinc could also be recovered.

The 1929–32 depression also hit Peru. Operations in Colquijirca were temporarily shut down in 1932 and stayed shut for the next few years.

Fernandini left the mine in charge of a young mining engineer as caretaker – Ernesto Baertl Schutz (see pages 95–6).

During the remaining 15 years of his life, until his death in 1947, old Don Eulogio Fernandini de la Quintana expanded the fortune and economic base of the Fernandini family still further. Over the years, he bought land around Lima, which, eventually, as the city expanded, became urbanized and valuable. And quite by accident – because his company was accused of polluting the sheep-grazing lands near the smelter and he had to indemnify the owners or buy them out – he acquired quite a large areas of land in the central sierra, which he made into the largest sheep farm in Peru. He also bought large tracts of agricultural land along the coast, which was first used for sugar cane growing and then for cotton.

Finally, he also brought Santa Barbara, the well-known Peruvian mercury mine, back into operation. The general story of that mine has already been told. The involvement of the Fernandinis in the Santa Barbara mercury mine dates back to about 1915. We saw how the mine was abandoned during the times of Peru's independence movement. Throughout the nineteenth century, work was attempted, off and on, but never with meaningful results. Even general, and occasional president-of-the-country, Luis Cáceres owned the mine for a while. He tried to work it, but abandoned it.

By 1915 Eulogio Fernandini had become well known, successful and wealthy. A friend challenged him to try his luck at these abandoned workings, where many a good man's luck had failed. He accepted the challenge, picked up the claims, and began work. A tunnel was driven, somewhat below the assumed level of the old Spanish workings, but only very low grade ore was found. Fernandini gave up again, and another 30 years were to pass before the Fernandinis looked at the mine once more.

By around 1940, undoubtedly prodded by war demand and higher prices for metals, they tried again. They acquired a Gould plant, two rotary kilns and a 1000 kW hydroelectric power plant. When the furnaces were ready, but before the power plant could be set up, World War II was over, the price of mercury dropped substantially and work was once again stopped. One more period of abandonment followed, from 1945 to 1955. By then, two of the company's senior executives, Carlos Bellatin la Rosa and Gonzalo Lostonau, insisted that working the mine on a larger scale and as an open pit operation, with large earth-moving equipment, would make sense and should be tried. And it was done.

For the next 20 years, the old workings again saw some action and quite some substantial production. They mined ore with three pounds of mercury per tonne, the lowest grade cinnabar ore ever worked anywhere. With the total collapse of mercury consumption and its price in the early 1970s, the Santa Barbara operation could no longer be justified. Large scale industrial operations ceased in 1976 and since then only sporadic and small production has been maintained to keep the concession and to supply the limited demand of the domestic Peruvian market.

Antenor Rizo Patron and the rizopatronita

Books on geology list rizopatronita as one of the vanadium ores. It is named after Antenor Rizo Patron, a young engineer who, in 1905, was working for Eulogio Fernandini as superintendent of the Huaraucaca smelter. They were looking for coal, as the Huaraucaca smelter needed fuel, on the Hacienda Ragra near the Punrun lake.

One day, Rizo Patron found something that looked like asphaltite, a type of coal, but it did not burn, or did not burn right, and could therefore not really be coal. It had to be something else. It was analyzed and found to be vanadium. Antenor suggested to his boss and the owner of the hacienda, Eulogio Fernandini, that he, Antenor, go the next morning to the town of Cerro de Pasco, to register the mining claim. Fernandini's reply was clear: Antenor had better leave that same evening, to be able to file the claim first thing the next morning when the office opened. So it was done. Fernandini a few years later sold the claim to Vanadium Corporation of America, and the mine, and Peru, became, for a while, the largest vanadium producer in the world.

Today the mine is closed and abandoned and no more vanadium is produced in Peru. But Antenor Rizo Patron's name continues in the name of the type of ore he discovered.

The last twenty years

In the early 1990s, Peru continues as a major metal producer. In fact, its mining production shows the proud listing given in Table 6.2. But development during the last few decades has been uneven. The value of the mining products the country has exported reveals at one glance how mining production grew and then stagnated (Table 6.3).

We should take a look at a few specific metals.

Lead

Production of lead is today, 1991, about where it was 10 years ago; in fact, where it was also in about 1970, and only slightly above the level of 1960. The potential for production above 200 000 tonnes per year seems to be there, but something or other always happens to prevent it. Still, production topped 192 000 tonnes in 1989, reached 189 000 tonnes in 1990, and went up to 199 000 tonnes in 1991.

Table 6.2. Mining production figures for 1989

Metal	1989 production (tonnes)	Ranking in the Americas
Lead	192 000	4th
Copper	298 300	4th
Zinc	485 000	2nd
Cadmium	303	4th
Silver	1 551	3rd

Table 6.3. Mining production figures between 1950 and 1991

Year	$ million
1950	45
1968	454
1973	631
1978	530
1983	1629
1989	1641
1991	1700

Copper

The picture here is similar. In 1960/1, production jumped from 50 000 tonnes a year to 200 000 tonnes, as Toquepala came on stream. Then, in 1976/7, production took another leap, from 180 000 tonnes to 350 000 tonnes, as a result of the startup of Cuajone. For a few years, production then even topped 400 000 tonnes, for an all-time high of 406 000 tonnes in 1987. But by 1990, the quantity was again down to 317 700 tonnes, to then rebound to 381 000 tonnes in 1991.

Zinc

For zinc 1991 was to be a banner year, and when the numbers were added up, production indeed topped 628 000 tonnes, better than the best previous year, 1987, when 617 000 tonnes were produced.

Silver

At 1770 tonnes in 1991, production of silver was 50% above the 1980 and 1970 levels. Between 1969 and 1981, production had stagnated in a rather narrow band between 1100 and 1300 tonnes per year.[21]

Gold

Production of gold was, at 14.6 tonnes in 1990, double that of 1981. Much of it continues, no doubt, as a by-product of the complex lead, zinc and copper ores. But there were also increasing activities in the Amazon area – wash-gold from rivers and dry creek placers.

How much was the combined loot from Cajamarca and Cuzco – the values the conquistadors took? Eight tonnes of gold and 40 tonnes of silver? Such quantities are now produced in a few days or a few months.

The remarkable story of Perubar

Driving up, or down, the Carretera Central – the principal highway from Lima into the sierra (the road that leads to La Oroya, Cerro de Pasco, and most of the mines in the central mountains, and that runs parallel to the Rimac river, and to the Ferrocarril Central, the principal railway line of the country), the casual traveller would never notice a mining operation quite close to the highway, and on the steep slopes of the narrow valley. Along the winding road, the view of the mines is but for a few seconds – then it is another scenery, more rocks, more precipices, and all breathtaking.

But a mine there is, near the town of Matucana, about 50 kilometres from Lima. And it is a remarkable mine. It is, in fact, the second largest zinc concentrate producer of the country.

Back in the 1950s, the Revoredos discovered barite crude ore – from which drilling mud for the oil industry is made. In 1957, they formed Perubar SA, and then associated themselves with Baroid Corporation, a subsidiary of NL Industries which, in 1986, sold the Peruvian operation to Rich & Co of Zug, Switzerland. Rich & Co had decided to invest in Peru at a time when most other foreign groups withdrew, or tried to withdraw, from the country.

The original, small barite mine – adits and handpicking – had become, in 1968, an open pit operation. And, in 1979, after increasing 'contamination' of the barite ore with zinc produced problems, a flotation plant was added, to henceforth produce crude barite ore, barite concentrate and zinc concentrate. In 1982, an adjoining (underground) ore body was opened up which contained principally zinc but also small quantities of lead and silver.

Rich & Co's investment paid off. By 1991, Perubar was one of only three or four profitable producers in the mining sector (all others are apparently losing money). The 1991 production reached 106 000 tonnes of zinc concentrates, 11 000 tonnes of lead concentrates and 48 000 tonnes of crude barite ore.

A look into the future

Shall we dare to look into the future? Peru was, is, and remains a mining country. Mining always played an important role in the development of the country. New mines were always discovered, and there are, beyond a shadow of a doubt, hundreds more new mining prospects known and waiting to be opened up. The long mining tradition of the country has produced a professional class of entrepreneurs, engineers, geologists, metallurgists, and other specialists, and also tens of thousands of miners. The future of this industry must thus be bright.

Admittedly, the problems the country faces at the moment appear exasperating. The Shining Path insurrection has already gone on for 10 years. The prevailing economic chaos, the origins of which are traceable directly back to the populist, socialist and xenophobic policies of the Velasco Alvarado dictatorship (1968–75) is far from over. Alan Garcia (1985–90) was another populist who could not leave the economy alone and thought he could solve Peru's problems by inciting labour, fiddling unrealistically with the exchange rate, and milking the mining companies more and more. It is a miracle that the mining industry has not only survived the setbacks and shackles of the last 20 years, but has even managed to grow in a few areas. The Peruvian mining community, and the country, deserve better. Maybe the government of Alberto Fujimori which took office in 1990 will finally bring some progress again.

Yet it is not only the turmoil inside Peru that conspires against the industry. International markets have also changed. Prices and demand are low, and the economics of mining are simply not positive for a country like Peru which depends on production from complex ores and deep, underground mining.

The 1991 annual report of the Peruvian Mining Society, Sociedad Nacional de Mineria, makes for dismal reading. The report produces a summary according to which only 5 of the top 18 companies made money in 1991. The company which made most money was the Southern Peru Copper Corporation, working the two open pit copper mines of Toquepala and Cuajone. Orcopampa and Perubar also operated profitably. Centromin, the former Cerro de Pasco operation, showed the largest loss. The industry was, during the early 1990s, obviously in a major crisis.

Notes

1. Pike, *The Modern History of Peru*, page 7.
2. Ibid, page 8.
3. Ibid, page 14.
4. The population estimates correspond to 1990.
5. I have chosen to use the English spelling of Bartolomew. In Spanish, Medina would have been Bartolome. But the Welser, a German, would have been Bartolomeus.
6. News, especially important news, travelled fast, even in those years, and it is quite plausible that the news – and physical mercury – took but a year or two to get from Mexico to Peru.
7. This is what the chroniclers and historians say. Although I am more than willing to go along with this concept, I have some doubts. Especially after the production

of mercury at Santa Barbara became a crown privilege, some kind of tight control must have existed, especially as no mercury was consumed at the mine and most was transported to Potosí. Just one or two roadblocks somewhere, and almost all mercury would have been taxed.

8. Most of my material comes from the old El Brocal company, owned until a few years ago by the heirs of Eulogio Fernandini. Members of the family and former executives let me see old records, took me to the mine, and helped me to understand the past. I am deeply grateful to them. I also owe much to Guillermo Lohmann Villena, *Las Minas de Huancavelica en los Siglos XVI y XVII*, as well as to WFC Purser, *Metal Mining in Peru*. See comments on Purser and his book in the note 2 to Chapter 1.
9. Purser, page 58.
10. Palma, *Tradiciones Peruanas*, page 422.
11. The Cerro de Pasco story comes from personal recollections of what executives of the firm told me in the period 1965–75, also innumerable pamphlets put out by the firm, newspaper articles in Lima during those years, and also, again, Mr Purser's excellent book.
12. Put together from *Ingeniero Andino*, July 1977, from pamphlets of the firm and from newspaper articles.
13. *Ingeniero Andino*, July 1977.
14. Ibid.
15. The author knew Alessio Sibille during the period 1959–74, and the story related here reflects what he told me.
16. Ernesto Baertl Montori and his mother agreed to be interviewed in June of 1979, in Lima, and related many of the details contained in this story. The author also knew Ernesto Baertl Schutz as well as Aquiles Venegas, Axel Nicander, Luis Picasso, Leon Rosenshine, Cy Fleischman, Carlos Montori and others of the group.
17. The author knew both Felipe Zacarias Montañez and Felipe Zacarias Alvarez.
18. From Langdon, *Don Roberto's Daughter*.
19. The author knew several executives of Atacocha, including Gerardo Diez Gallo, Edgardo Portaro, Francisco (Paco) Gallo and Hernando Lavalle.
20. See also note 8 above. The author knew many of the members of the Fernandini families, Jorge Pablo Fernandini and his wife, Iris Fernandini de Ravago and her husband Enrique Ravago, Anita Fernandini de Naranjo and several of her children, as well as several executives of the group, notably Carlos Bellatin, Gonzalo Lostonau and Alberto Rubina.
21. The statistics come from *Carta Minera – Anuario 1990*, published by the Andean Mail & Peruvian Times SA, an excellent publishing house in Lima which has, over the years, maintained steady high quality reporting on the Peruvian scene, including the mining industry. I have also relied on the annual *Metallstatistiken* put out by Metallgesellschaft Aktiengesellschaft.

CHAPTER 7

Bolivia

Mining in Bolivia did not begin with Potosí, nor did it end with Patiño, but these two are probably the outstanding milestones in five centuries of mining history filled with remarkable events. Although Mexico, Peru and Chile, as well as Brazil, have mining industries that played, and play, very important roles in the growth of these countries, it is Bolivia that was and is South America's mining country *par excellence.*

In Bolivia, the origins of mining and metal working go back to pre-Columbian days. The early inhabitants, even those that predated the advent of the Inca empire, knew and worked gold and silver, copper and tin. And they also knew bronze, the alloy of copper and tin. They worked it to an exceptional degree of hardness, so hard in fact, it was reported, that they cut, or chipped, rock with it. We do not know whether they found copper and tin separately, or just found them together, naturally bound or alloyed in the form of native bronze, as they found copper, silver and gold. They called their bronze 'champi', and its exceptional hardness was no doubt obtained by hammering it into shape with rocks; by trial and error, or by accident, they found out that the more they hammered, the harder the metal became.[1]

What was said in Chapter 6 about Peru also applies to Bolivia. The two countries were in pre-Columbian times one and the same, in terms of mining and metal working technology. Here also gold and silver had, prior to the arrival of the Spaniards, no value other than as property of the Incas and as adornment for the ruling class. Gold and silver were then manifestations of the sun and the moon. Copper and bronze had value for the making of tools.

The conquest of Peru by the Spaniards took place – at least the most dramatic early phase – between 1532 and 1540. Once the area of what is modern Peru was more or less under control, the conquistadors moved into what the crown came to call 'Upper Peru', which later was to become Bolivia. They also moved into Chile, northern Argentina, and, as we already saw, Colombia and Equador.

The Fuggers

Bolivia, or 'Upper Peru', as well as much more of southern South America, came close to being settled and administered by the Germans.

We already saw how the German banking house of Welser obtained a grant to explore and exploit Venezuela, in payment of IOUs the Emperor Charles V owed (see page 55). When the other big banking house, Fugger, much more powerful and more prominent than the Welser, heard of this deal, it also wanted its share. Negotiations began.

In 1529, Charles V had made an agreement with Francisco Pizarro, who subsequently set out to conquer Peru, that Pizarro's rights would go from Tumbes, at today's border between Equador and Peru, south to where today is Chincha. The region south of Chincha was signed away to the Fuggers on 28 June 1531, as far south as it went and 200 Spanish leagues (of 5572.7 metres) inland. Thus, all of western South America, to the south of today's Peru, including Bolivia and Chile, was signed over to the Germans – but at first only for eight years of exploitation.

The Fuggers never got there. A major expedition was lost in Patagonia, or in the southern Atlantic. The fleet was literally 'lost'. The rights of the Fuggers never took effect as they apparently made no efforts to put together another expedition.

When, a few years later, Diego de Almagro, one of Pizarro's companions who felt he had been cheated out of his fair share of the conquest, wanted the region, the crown gave it to him. It probably appeared prudent to Charles V to try to placate Almagro and thus to avoid a civil war between Almagro and Pizarro. The crown assigned the lands a second time, hoping that time and circumstances would solve the matter. Time indeed did. Almagro soon met a cruel death and, consequently, these huge southern tracts of land reverted to the crown. It was not the Germans, nor Almagro's heirs, but others, including the crown, who cashed in on the silver of Potosí.[2]

Charles V and the Spanish crown were great 'givers', especially of what was not theirs to begin with!

The silver of Potosí

The conquistadors did not get much loot in the form of adornments and jewellery in Bolivia – at least in the beginning. But a few years after the

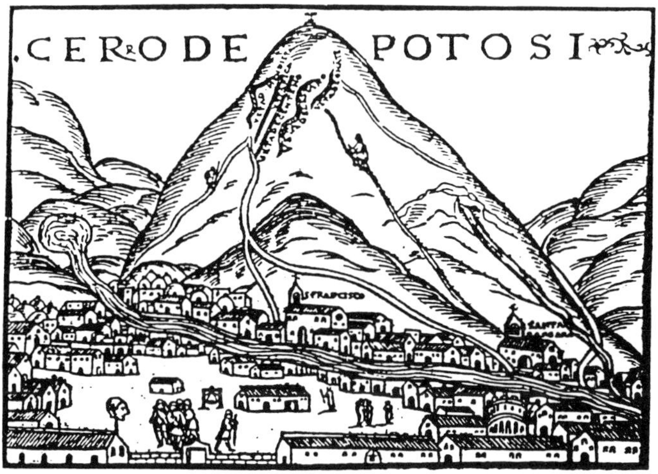

Two views of the Potosí silver mountain. Sources: the first illustration is from *Cronica del Peru*, by Pedro Cieza de Leon, published in 1553 and the second is from Johan Ludwig Gottfriedt, *Newe Welt und Americanische Historien*, first published in 1631 by Matthaeus Merian dA and republished in 1980 as Stuttgarter Faksimile Edition im Fackelverlag, Stuttgart, Germany.

conquest, in 1545, a spectacular discovery was made in the barren plateau of Bolivia, some 4200 metres above sea level, above which towered yet another thousand metres of mountain. It came to be called Potosí.

After Almagro's death, the Pizarro brothers took over the area that had been under his control. But then Francisco Pizarro was also killed. And anyway, the area around Potosí was not considered of value during those years, and nobody really cared who owned it.

One day, so the story goes, a native shepherd by the name of Huallpa spent the night with his flock on the slope of the Potosí mountain. In order not to lose the animals, he tied them to some shrubs. The animals pulled several shrubs off the rocks, and in the morning Huallpa beheld a very rich, very metallic-looking vein. It must have been unusually wide and pure.[3] It turned out to be silver. Huallpa told his master, one Juan Villaroel who, on 21 April 1545, filed the first mining claim.

We should not take the story of how Potosí was found as the absolute truth. In other versions, the shepherd built a fire and soon rivulets of liquid silver ran down the mountain. Whichever way it was, the shepherd got little, or nothing, of the wealth.

Villaroel told Diego Centeno, justice of the city of La Plata – also known as Chuquisaca and today called Sucre – who immediately sent some 60 Spaniards to start a settlement and work the mine. Soon over 5000 holes and adits existed.[4]

The wealth the early miners pulled out of the Potosí mountain must have been simply fabulous. One of the early chroniclers, Francisco Lopez de Gomara, wrote that the ore assayed 50% silver and more. The historian Antonio de Herrera claimed that not even a third of the silver mined in the beginning was registered. Crown law required that all gold and silver had to be registered with the royal authorities, to then be coined. This meant that the Royal Fifth, the principal tax, and the registration fee had to be paid. The first mint in South America[5] had been set up in Lima in 1536, but it was not until 1565 that good equipment arrived. Eventually, in 1577, part of the installation and equipment was transferred to Potosí by order of Viceroy Toledo. Thus, between 1545 and 1577, the production of Potosí had to be taken to Lima for registration. It is reasonable to assume that only a trickle of the total reached Lima.

By 1548, 13 scant years after the discovery, the richest veins began to peter out. Although in about 1562 new discoveries brought renewed bonanzas and silver 'rushes', veins of near pure silver were something of the past. Narrower and narrower were the veins they had to mine. But there was also the lower grade ore – what could been done with that?[6]

The Spaniards in Lima and in Potosí had by then heard of the patio process and amalgamation. By the late 1550s, some mercury was imported from Almaden in Spain. Transporting this liquid metal all the way from Spain to Potosí must have been quite a task. Leather bags were used – and the results were not satisfactory. Much material filtered out through the seams into the holds of the ships, and during the transport by mule across the isthmus of Panama, and again during the onward voyage down the west coast of South America to Callao, or Pisco, or Arica, or wherever they discharged the cargo for Potosí in those years. Still more mercury was lost during the final – and lengthy – trip by mule or llama into the mountains to the final destination. At the very end, little of the precious liquid arrived. And, as the miners were not certain of the process, and had to experiment, they were not very successful. It took a few years before the Mexican version of the patio process was adapted to the different ore and conditions on the Bolivian high plateau.

In about 1550, instructions must have been given by the crown to the viceroy in Peru to look for mercury mines. Town criers made the news public.[7] Then, in 1563, Amador Cabrera found – or was led to – the cinnabar deposit near Huancavelica.

We do not know from where, or why, Cabrera knew of the value of the orange-coloured cinnabar ore. The story is that he sent samples to Lima where they were seen by one Enrique Garcés who had worked in Almaden and who immediately recognized the typical colouration. Another version of the story is that Garcés was shown the sample and pronounced it anything but cinnabar ore, as the sample was of a dark orange rather than the light orange tone that he remembered from Spain. However, they assayed it and it was the mercury-containing cinnabar ore.[8]

Potosí now had a new lease of life, with large quantities of mercury coming from a nearby source.

The history books are full of stories about what wealth was extracted from the Potosí mountain. The calculations are very complicated and the reader is warned that whatever figures are shown are only approximations as the old Spaniards used many monetary units – as was already discussed in connection with the Caribbean (page 23) – such as marks, *pesos corrientes* (current pesos), *pesos ensayados* (assayed pesos), reales, ducats, and 'pieces of eight'. Also, a gold mark had 50 pesos, but a silver mark only 5. I shall make a few calculations and leave the fabulous conclusions to the reader's imagination.[9]

The production and tax data for the early years are very sketchy but, in 1549, the tax supposedly came to 25 000/40 000 pesos per week or, at 27.548 grams per silver peso, about 700–1000 kilograms per week. In the three years 1548–50, the total recorded production supposedly reached 15 million ducats,[10] or some 384 tonnes. One historian, Ballantine Cobb, gives figures that the crown received about 1 million pesos per year as Fifth for the period 1548–57 which means that the yearly production of silver must have been about 5 million ducats per year, or about 135 tonnes per year.[11]

The richest of the original veins were 'white' silver and often finger thick.

Mining in the sixteenth century: breaking pieces of ore with iron hammers (although this scene is from Germany, the same methods and tools were applied in the Americas). Source: Agricola, sixteenth century.

The miners just concentrated on the highest grade ore and dumped the rest outside. In the beginning, they only worked the rich oxide (oxidized) ore at or near the surface. The equally rich, but blackish in colour, sulphide ores were disregarded.

When the rich oxide ores from the surface ceased to be available, in the 1560s, the process of amalgamation was applied to the sulphide ores and production substantially picked up again from about 1575. By about the year 1600, some 150 ore treatment plants existed. By 1768, their number had again shrunk to 25.[12]

Orlando Capriles Villazon, a Bolivian historian, refers to tabulations that show a total registered extraction of about 1.6 billion ounces between 1714 and 1800.[13] This would give about 50 000 tonnes – yes, fifty thousand tonnes of silver! Two other Bolivians, Manuel Cortéz and Modesto Oriste, calculated an extraction of about 1650 million pesos in only 11 years, between 1545 and 1556 – or about 45 000 tonnes.[14] Yet another Bolivian,[15] Luis Peñalosa, calculated for the entire period 1545 to 1800, an extraction of about 2.5 billion ounces, or about 69 000 tonnes. One of the more original, or primary – but not necessarily more reliable – sources, Dr Pedro Vincente Cañete,[16] who was governor of Potosí in 1797 and who had available to him many of the original sources, tabulated 820 million pesos, or about 22 000 tonnes, produced by 1783, but he counted only what was recorded in the official records. How much was smuggled out? Many historians claim at least half of the total production.

In the final analysis, and when all the books and records are researched and tabulated, does the exact value really matter? Were we indeed to agree that the value of all the silver extracted from the Potosí mountain throughout three centuries amounted to $10 billion, what would it mean? Poor Bolivia, and poor Peru – as Potosí was then part of Peru. Little of the wealth stayed; all the value left the country. Neither country benefited, developed, or progressed in those years. Even the taxes paid at the source went into the hands of the crown – most to be sent to Spain, little to be paid to the officials in residence. The rest of the wealth stayed with the mine owners, also Spaniards, who squandered it; a few of them returned home with their take.

It is obvious that both Peru and Bolivia have been catching up only now, in the twentieth century. The Spaniards plundered these countries for 300 years or so. It is fashionable today to accuse the Yankees of having plundered South America in recent times, but the facts of Spanish colonial history make the Spaniards the real villains; whoever came after them were imitators on a much smaller scale.

The importance of this unbelievable wealth for Spain and Europe should once more be stressed. In the sixteenth century, Spain was still a backward country. The changes of the Renaissance and of the Counter-Reformation had not yet touched it. Spain continued very much medieval and feudal. The crown, the Catholic church and the large landowners ran the country. Its economic foundation was agriculture. Its social pyramid was based on serfs at the bottom and parasitic lords at the top. The voyages of discovery and the conquest and settlement of the Americas were necessary and welcome escape valves for some of Spain's pressing social problems. The adventurous young – and some not so young – picked themselves up and went overseas. Many of the poor also followed. To sail to the Indies meant hope, opportunity, excitement, and for some eventually wealth and fame.

The rich stayed behind, maintained the feudal order, enjoyed the riches and traded with the rest of Europe. But many of their sons left for a new life in the Americas.

The wealth which came about in Potosí and Lima created demand for manufactured products – luxury items – from the home country. The miners – and the small scale merchants who set themselves up – suddenly had money. They demanded, in ever increasing quantities, imported goods such as tools and weapons, clothes and utensils, furniture and other amenities; and necessities they remembered from home. They also wanted better housing.

The sudden surge of such demand in the mining centres had outward ripple effects. The demand for goods flowed back to Lima, on to Panama, to Spain, and then spread all over Europe. And when the goods began to move toward Potosí, many were the hands busy in between the suppliers and the final consumers. The wealth of Potosí thus caused chain reactions. Traders, builders, exporters, importers and shipowners benefited. Many became wealthy. And because of this new wealth, still more demand was created.

The focal point for the new trade with the Americas was Seville, Spain's overseas exit and entry port. From Seville, the demand for goods spread to other parts of Spain and Europe, as did the wealth.

Production of silver at Potosí remained fairly stable throughout the seventeenth century but began to decline during the eighteenth. In 1776, as we have seen (page 80), the Huancavelica mercury mine collapsed. Production of mercury thereafter continued only on a small scale – and this naturally affected production of silver at Potosí.

By 1827, Potosí was a mere ghost town with only about 9000 inhabitants, down from a high of about 160 000 people in 1610 when Viceroy Montesclaros had the people counted.[17] By 1975, Potosí was back up at 75 000 inhabitants, and the number today is probably around 100 000 people.

But back then, in around 1552, an egg cost half an ounce of silver, a chicken 6 ounces, a litre of wine went for 1 kilogram of silver and a yard of brocade for half a kilo of the metal.[18] And in its heyday, Potosí had 360 bars, 8 fencing schools, 36 gambling casinos, 4 theatres,[19] even 36 churches, and many, many bawdy houses.

The first half of the nineteenth century saw Bolivia struggling to shake off Spanish colonial control, which culminated in independence in 1825, and struggling thereafter to overcome a never ending sequence of internal strife and chaos. Not much mining took place during this period. It was only during the second half of the century that meaningful production began again.

The Aramayos

The Aramayos trace their family back to Navarre and the early seventeenth century. In 1639, one Diego Ortiz de Aramayo went to Bolivia where, together with a son, a cousin and a nephew he appears to have been owning mines by 1691. But it was to be only in the nineteenth century that the family came to wealth and power.[20]

José Avelino Ortiz de Aramayo (1809–82) began, in 1824, as a cattle driver and general trader travelling between northern Argentina and Bolivia. A few years later he began clerking in a store in Tupiza. As his boss also owned a few mines, Aramayo became familiar with what mining meant.

In 1835, a then well-known and well-off miner, Guillermo Irigoyen, hired him as secretary and took him along to Europe. The trip, via Buenos Aires, took three months. In France and England, Aramayo's natural curiosity caused him to notice the latest technological inventions, among them steam engines and locomotives as well as a primitive mining railroad. He bought, for his own account, general merchandise which, back in Bolivia, he managed to sell at considerable profit. Two years later, and because of the success of the previous trip, he went back to Europe. This time he took along some gold ore which he also managed to sell at a profit. By 1845, he was a merchant in general goods, a trader in minerals, and about to become a miner. He next went to Chile to help his father-in-law who owned several mines there.

Moving constantly around, he came to know northern Chile and southern Bolivia intimately and became more and more convinced that mining could be a very lucrative business. He therefore formed, in 1849, together with several brothers, a company called Aramayo Hermanos which was to be active in general commerce and ore-buying. A year later, he owned his first mine, called Carguaicollo, an old silver property. He now put to use what he had seen in Europe; he mechanized the mine, introducing an underground railroad – small cars pushed manually along wooden tracks covered with iron sheeting. He also imported up-to-date milling equipment and built a concentration plant as well as a small refinery.

By 1858, Aramayo was sufficiently encouraged, and in funds, to expand. He formed Compañía del Reál Socavón de Potosí which was meant to tackle once again some of the old and abandoned workings of the Potosí mountain. He wanted to apply up-to-date technology to go into parts of the mine and to work low grade ores that the Spaniards with their primitive methods had not been able to handle in colonial times. As the expenses exceeded the resources of the firm, Aramayo issued new shares and borrowed liberally. And right away he went after other mines, including a large and well-known one: Huanchaca. They were all silver mines. Aramayo also realized, and was probably the first to do so, that Bolivia had a great potential in tin.

But the Reál Socavón de Potosí did not bring him wealth and he also had to abandon Huanchaca which, but a few years later, would bring wealth, fame and eventually even the presidency of the country to another miner, Aniceto Arce.

By 1864, Aramayo Hermanos was bankrupt. José Avelino had, however, saved some assets and began all over again, but now assisted by his son, Felix Avelino Ortiz de Aramayo (1846–1929). By the early 1870s they were apparently well enough off again for the father to afford to establish himself in

Paris where, forbidden for health reasons to return to the altitude in Bolivia, he spent the rest of his life. But the family fortunes fluctuated as he apparently left his son more debts than assets when he died in 1882.

Although Felix Avelino was born in Paris and had received a thorough European education, he began to work as early as age 16, as cashier in one of the mining camps in Bolivia. But this did not last long and he was soon back in Europe for some more schooling, including an apprenticeship in a tin mine in Cornwall. By 1866 he was back in Bolivia. He now formed a new company, Felix Avelino Aramayo y Cia; two German chemists by the name of Francke were his partners.

He soon got hold of a bismuth mine and expanded it. Only a year later he travelled to London with half a ton of bismuth metal. The business was so profitable, apparently, that he even set up a small refinery in London which began operating in 1868. While in Europe, he bought mercury and shipped it to Bolivia for the recovery of silver. But for quite some time to come, bismuth was to be the most profitable part of the business.

By the 1890s, the silver market began to recover and the silver mines of Bolivia (as those elsewhere), among them those of Aramayo, finally began to make some money again. And as from about 1900, tin became important, contributing substantially to his income. By then he had also acquired several wolfram mines.

For a while, in the early years of this century, Aramayo must have been the largest and the wealthiest of the miners in Bolivia. But Patiño now quickly surpassed him and by about 1925, Hochschild became the second largest.

With mining production having represented in some decades 80%, 90%, and even more of Bolivia's exports, and, through export royalties, of its government's income, it is easy to see how large an important miner such as Aramayo looms in the history and the development of the country. Taken together, the principal mining tycoons really constituted the backbone of the country.

Carlos Victor Aramayo was the third of the family to play a significant role in the Bolivian mining scene. He was also born in Paris, and was raised principally abroad. It was only at age 22, in 1911, that he went to live in Bolivia. He began to work as secretary to the manager of one of the family mines. But he soon took on higher positions and in time assumed control of the family enterprises. He also entered politics inside Bolivia and, as a diplomat, represented Bolivia in various positions abroad. By 1925 he became president of the new holding company Cie Aramayo des Mines en Bolivie, domiciled in Geneva.

When, during the Chaco War between Bolivia and Paraguay, 1932–5, the government appealed to the mining industry to contribute to the war effort, Aramayo signed up for 10% of the total amount from the industry. By comparison, Patiño contributed 64% and Hochschild 20%. As the individual shares must have been based on some factual schedule, we can consider these percentages as reasonably reflecting the relative strength of the different mining groups.

The 1930s were overshadowed by the depression, the collapse of the tin price – and other prices – in 1929, the very slow economic recovery in the following years, the results of the First International Tin Committee (see page 126) and the approaching clouds of World War II. By the late 1930s, the miners were again making money, but strict price controls during World War II prevented the producers from recovering after the bad years of the 1930s. The necessary and indispensable reinvestment and modernization did not take place. Social unrest after the war foreshadowed the events of 1952 when the government expropriated the large mines, and the Aramayo group, for all practical purposes, ended its involvement in the mining industry of Bolivia.

Patiño and the tin mountain

Whereas Aramayo was the first of the big mining men in the modern history of Bolivia, the second, Simón Iturre Patiño, was destined to become, in the early part of the twentieth century, the wealthiest and most powerful of them. It was his spectacular success and wide-ranging international involvement which gave rise to descriptions of him such as 'tin king' and 'tin baron'. These he undoubtedly was, a giant in terms of riches, industrial power and political influence. But these descriptions, which are also applied to Hochschild and Aramayo, are often used today with derogatory connotations, which is one-sided as it deliberately ignores the positive contributions of these men to the development of the tin mining and consuming industries, as well as to Bolivia in general.

Patiño was born in 1860 in a village near Cochabamba.[21] His roots, at least on his father's side, were Spanish Basque. He was to reach the age of 86 before he died, in 1947, in Buenos Aires. In between were 40 years of hard and stubborn work, search and slow growth (1860–1900), spectacular success and instant wealth as well as rise to international prominence during the next quarter century (1901–23) and, following a heart attack in 1923 which forced him to leave high altitude, the patient building and administration of a growing international organization based in Europe and the United States. After 1923 he never returned to Bolivia.

Neither his education, through high school in Cochabamba, nor his first jobs as a salesman for a small importer in Oruro and then as a clerk in the Huanchaca mine, foretold his later success. We already saw that the Huanchaca mine had been opened again by Aramayo but had then gone into the hands of Arce who, together with Chilean capitalists, brought it to a new boom period between 1880 and 1900. In 1894, Huanchaca paid dividends amounting to £400 000, a sum which about equalled the total income of the Bolivian government that year.

During Arce's presidency, in 1892, the railroad between Antofagasta and Oruro was completed. In that year we find Patiño back in Oruro, now as an employee of Herrmann Fricke y Cia, a commercial house dealing in general merchandise, but also acting as buyer of cinchona bark (the raw material for quinine acid), rubber and minerals. This is where Patiño learned the procedures of purchasing and shipping ores. It was his first contact with the *rescate* business (see pages 127–8).

He became acquainted with one of the small miners, Sergio Oporto, the owner of a 10 acre concession called La Salvadora, on the Espirito Santo hill, near a town called Llallagua. There he produced small quantities of complex and low grade tin ores. Oporto was not very successful, needed additional working capital and, in 1895, invited Patiño to become his partner. Patiño accepted. He continued to work with Fricke but now also acted as seller of the production of the Salvadora mine; Oporto stayed at the workings. However, they continued without success. In 1897, Oporto wanted out, and Patiño became the 90% owner of the mine. The remaining 10% was in the hands of Fricke.

Patiño now resigned his job at Fricke and began to look after the mine full time.

Work there was hard, with the camp located at 4850 metres above sea level. There were several moderately successful mines on the lower slopes of the mountain, among them a group of mines called Llallagua. But near the top, Patiño was alone. He had no power, no machinery, only hand drills, hammers and pickaxes. He hand-worked a stone rocker to crush the ore; it was then hand-separated, bagged and sent down the hill on llamas, for a three or four day trip to the nearest railroad station. At the beginning, Patiño cannot have had more than a handful of labourers, perhaps at most 10. And he worked as one of them, not content to just direct the work. But he also handled the sale of the ore, the logistics of transporting the ore down and the purchasing and transporting of supplies back up.

One of his fundamental, and no doubt intuitive, decisions was, in 1898, when work in a gallery ended in worthless pyrite, to follow the pyrite nevertheless. At this time, the mine had never yet paid. He had run up a large debt with Fricke, was just a few weeks away from no longer being able to pay his staff, and was close to giving up. On one of those occasions of despair, his wife – he had married in 1889 – sold some old family jewellery to shore up the accounts. And he plodded on, guided by stubbornness and blind faith that he was yet going to hit the rich vein soon. Fricke continued his support, with money and supplies.

And then Patiño struck the vein. One day, in 1901, the foreman came running with some pieces of shining, metallic-looking ore. He thought it was solid silver, but Patiño suspected it was tin. And it was. The samples assayed 47–57% tin, the richest vein in the region.

Continuing for a while with the precarious and primitive methods used so far, he nevertheless managed to pay off most of his debts. And he had to fight off attacks by armed competitors who, envious of his success and using at best dubious counterclaims on the property, tried to expel him and his people. But he prevailed, increased his staff, and the vein continued wide and rich.

By May 1902, his production was already at 120 tonnes a month. His formal style was now 'Empresa Minera La Salvadora, Simón Patiño, Uncia'.

Ore sales and legal defence became of overriding importance and he moved to Oruro. The mine was operated by competent staff. He next rounded off his property by registering some marginal acreage adjacent to La Salvadora.

By late 1905, Patiño managed to pay off the last of his debts to Fricke and he bought out Fricke's 10% holding in the mine. He was now finally the sole owner. He also stopped selling locally to Fricke and began shipping his ore to Avelino Aramayo & Co in London. And – the surest sign of his phenomenal success – he bought an aerial rope-way in Germany to facilitate the transport of the ore down the hill. He also bought a leaching plant and began to import jute bags directly from Europe.

He was now making money hand over fist; it did not take long and he deposited over $100 000 with a dealer in machinery in Santiago, as downpayment for a big new mill which he had ordered from Europe.

The Espirito Santo hill was by now known as Salvadora hill. And although La Salvadora had for a while been the largest and best mine on it, another company now competed for top honours, an equally successful operator called Cia Estanifera de Llallagua, a Chilean firm mostly owned by investors from Santiago.

Patiño began to realize that too much of his profit depended on getting the right price from the smelters in Europe. His next step, in 1911, was to set up an office in Hamburg. Most of the smelters for tin ores were, in those years, in Germany and in England. The United States was to see a tin smelter only briefly, during World War I at Perth Amboy and again, from 1940 onwards, at Texas City. Patiño at once went a step further than merely selling his own production; he offered the facilities of his Hamburg office to act as sales channel for other producers.

Back in Bolivia, he now bought out two smaller neighbours, Cia Minera Uncia, and Bebin Hermanos, as well as a few other nearby mines. He also bought, in 1912, a half interest in Zinnwerke Wilhelmsburg GmbH. When he saw the war clouds were getting darker, he transferred his office to Paris.

During World War I, his sympathies were with the Allies. He even bought, in 1916, an interest in the UK smelter of Williams, Harvey & Co Ltd. As more and more tin was required during the war, and prices rose, he increased capacity at the mines.

This text reads like a repetitious litany: Patiño making money, increasing capacity, buying additional mines, participating in smelters, increasing capacity, making more money, buying more mines, etc. But it was a real success story, unashamedly lived and here retold. And all this in a mere 15–20 years.

By now, at the end of World War I, Patiño was approaching the age of 60. In 1923, he suffered a heart attack in Bolivia; medical advice was that he had to leave high altitude, and to stay henceforth at sea level. This meant he had to leave Bolivia. He once more sailed for Europe, never to return. But there was still one further and major step which needed to be taken in South America to round up his holdings: to acquire control of the Llallagua company.

Already in 1915, Patiño had begun, quietly and through third parties, to buy up shares of Llallagua on the Santiago Stock Exchange. He achieved his objective by 1924 and with control over Llallagua at once transferred the company to New York, incorporated it as Patiño Mines and Enterprises Consolidated, Inc and merged La Salvadora into it.

There seemed to be no limit to his appetite to gobble up more and more properties. Late in the 1920s he bought into the Cornish Tin Smelting Co Ltd, and Penpoll Tin Smelting Co Ltd, operating in England, as well as into the Eastern Smelting Co Ltd, operating in Penang, Malaya. In 1929, he set up a new company, Consolidated Tin Smelters Ltd, which henceforth acted as a holding for his smelting interests. And he continued to buy mines in Bolivia, reaching a point where, in 1929, he accounted for over half of Bolivia's tin output.

The 1929 depression hit the tin-producing industry badly. Consumption and prices slumped, and stocks piled up. The producers got together and formed, in 1931, the First International Tin Committee as a vehicle to 'continuously adjust production to consumption'. The objective was to stabilize prices at a level acceptable to both producers and consumers. A buffer stock was set up and quotas were assigned. The committee and its actions contributed to prevent a collapse of the industry and caused prices to begin slowly to rise again.

During World War II, Patiño moved to New York. The war effort again required production increases, but prices were now controlled, as they had not been during World War I. Bolivia's production of tin, at that point about 25% of total world output, had to make up, together with that of a few other countries, for the shortfall of tin from Malaya and Indonesia which had fallen into Japanese hands. Aside from tin, Bolivia also produced tungsten, lead, antimony and bismuth for the war, and considerable quantities of these also came from Patiño's mines.

The last part of the Patiño story came in 1952, five years after Simón Patiño's death, when the big mines were expropriated. With it, the involvement of the Patiños in Bolivia ended.

Mauricio Hochschild

Mauricio Hochschild was the third of the big mining men of Bolivia during the twentieth century. Contrary to Aramayo and Patiño, who were Bolivians, Hochschild was a foreigner.

He was born in 1881 in Biblis, near Frankfurt, Germany. He was a cousin of Zacharias Hochschild, one of the founders of Metallgesellschaft, as well as of the Hochschilds of the American Metal Company. And Siegfried Bendheim, of Philipp Brothers, was his second cousin.

He took a degree at the mining academy at Freiberg, Saxony, and began to work for Metallgesellschaft for whom he went, in 1906, to Australia to run a mine. When the venture was abandoned a few years later because of a prolonged absence of rain, he followed, in 1911, the suggestion of an uncle and travelled to Chile to take a look at business possibilities in mining. It is likely that he visited Bolivia then and learned first-hand of the spectacular success story of Patiño.

In 1914 he returned to Germany and spent the war years running a mine in Austria, but by 1919 he was back in South America and now began to buy ores – the *rescate* business.

Rescate is a Spanish term that goes back to colonial times; it derives from the word *rescatar*, meaning to recover. As early as 1776, a Banco de Rescate de Minerales was set up in Potosí, and it probably only institutionalized an activity which had commenced much earlier. Again, in 1850, Bolivian President Belzu founded yet another Banco Nacional de Rescate de Minerales.

Today, by far the major part of mining in the world is done in an efficient way, on an industrial scale. But until quite recently – actually until the advent of aerial prospection – mines were discovered by prospectors, farmers, shepherds and small scale miners. Until well into this century small mines accounted for a major share of total mining production. This was true for many countries, but especially so for the developing ones. In countries like Bolivia and Peru, for instance, such small scale mining was the only kind of mining that existed until the beginning of the twentieth century. It was usually just one man, with a few assistants, who worked a drift, sank a shaft, hand-picked ore, and shipped it out and down the mountains on muleback.

This is how the Incas mined, and this is the method the Spanish conquistadors continued to use after the conquest at the beginning of the sixteenth century. One notable exception to the *rescate* mining existed – the Santa Barbara cinnabar mine near Huancavelica in Peru which the Spanish crown tried, for a while, to work on a somewhat larger scale. But the attempt failed and small private operators continued to account for all mining until mechanization and the need for large scale investments brought big companies into Bolivia and Peru towards the end of the nineteenth century. But the small operators continued to play a major role until about 1960 when they began to disappear.

Rescate described the way handpicked ores were purchased in the interior and brought to market. Until about the year 1900, the only minerals worth mining were gold and silver. In a few places copper and lead began to give meaningful economic returns, but the quantities were small. Almost non-existent transport systems, technically primitive methods of mining and shortage of labour made larger scale mining very difficult. We should visualize transporting ores and metal by mule or llama trains for a week or two from the bleak Andean high plateaux down the slopes to the coast. But the most important impediment was the cost–price equation. Metals, other than gold and silver, were not very valuable until about World War II. Further, expensive production and transport costs were not compensated for by sales prices. Thus, those who did business in the other metals could not prosper.

The headlong industrial expansion of Europe and the United States during the second half of the nineteenth century changed much of this situation. More and more metal was required and the sources of supply in Europe were becoming exhausted. Also, values increased. And new, more sophisticated applications for metal were developed. In this process, miners learned to differentiate between more and more types of ores and metals.

This increasing demand projected backwards to the sources of supply, and the *rescate* business picked up. Itinerant buyers – originally travelling pedlars of all kinds of goods but in time specialists in metals and then ores – roamed through the mountains of the mining regions of Bolivia and Peru in search of ores to purchase – just as they had done in colonial times. Now, beginning in about the year 1900, general traders in the towns became the new ore and metal buyers who 'recovered' (*rescatar*) what the small miners produced. Now, it was not only gold and silver ore and metal, but it was also tin ore, and some antimony, bismuth and tungsten.

The miners brought a few mule or llama loads to town, and offered their product for sale or barter to the merchants. This is how companies like Duncan Fox, importers and traders in general merchandise, became involved in metal and ore marketing in Bolivia. Hochschild was apparently the first 'pure' ore buyer to establish himself.

In 1921, he opened an ore-buying office in Oruro. Offices in Potosí and Tupiza followed in 1923. He was successful and began to make money. As he travelled around and became familiar with the countryside and the history, he found that the lure of the Potosí mountain also affected him. So much wealth had come from it, it could not all be gone; more had to be in there. Many small and medium-sized operators still worked the old 'holes', still extracted something, and some still got wealthy. But instead of doing likewise, becoming one more of these small operators, Hochschild had the idea of joining forces, of getting a number of these small miners to form a company, naturally under his leadership, so that they could pool resources, avoid duplication of efforts, rationalize exploitation – and even reduce litigation. The Cia Unificada del Cerro de Potosí was the result. And the idea worked. Where until then many small companies had barely survived, Unificada now made money, and quickly became the second largest tin producer in the country – because the Potosí silver mountain also contained tin.

Another of Hochschild's early successes, and no doubt one of the cornerstones for his later fortune, was that he induced the German smelter Berzelius to develop a smelting process for low grade tin ores which until then could not be processed and were left as without any value.

Hochschild next acquired the Huanchaca mine; it had again been abandoned.[22] Applying modern methods, he succeeded in reworking the old colonial dumps from which quite some values were recovered. And the mine, pumped dry (it had filled with water) and now worked at greater depth, was once again a money-maker.

With fortune came power and he began to acquire mining properties and in time came to be included among the three most powerful mining magnates of Bolivia. He also acquired Cia Minera de Oruro and eventually opened up the Colquiri property.

Over the years, Mauricio Hochschild and his growing organization also branched out into Chile, Peru, Argentina and Brazil. For a while, in the 1940s and 1950s, the Hochschild group had a considerable economic impact in much of South America. Shortly before and during World War II, Hochschild was one of the havens for Jewish refugees from Europe. Many of them had nowhere to go and were lucky and happy to have got out of Germany. Hochschild offered them a job, as proof of employment was a prerequisite for a visa in most countries. But in the post-war period, he could not keep this highly intelligent and often over-qualified (admittedly in different professions,) group of immigrants together. Other companies, and among them Philipp Brothers, were to benefit.

The Hochschild mines were among those expropriated in 1952; this effectively ended Hochschild's participation in the Bolivian mining industry. He and his group were much vilified in the aftermath of the nationalization, in part, no doubt, especially so because he was a foreigner and a Jew. But, like Patiño, Hochschild was also a stubborn, imaginative, hard-working and often visionary builder. He revived the *rescate* business to where it became a major industry, and he opened up, or reopened, mines that others had left. As had Patiño, Hochschild profited. He died in 1965, at age 84.[23,24]

The 1952 nationalization

The large, wealthy, powerful and influential mines and mining companies in Bolivia were, as can be imagined, always the object of agitators, reformers and the envious. Their pay scales were always considered too low, never mind that there was in Bolivia, in general terms, no other industry worth mentioning, and their profits, real or just imagined, were always considered too high. The 1929 crash and its aftermath brought, as we saw, drastic reductions in price and then production; major hardship in the mines followed. And when tin prices improved again, wages did not rise in the same way. We also saw that the industry did not really see another boom period in the 1930s nor in the 1940s, and that the mines did not modernize.

Social unrest, labour agitation and political ferment grew during this time. In 1940, a new party was formed, the Movimiento Nacionalista Revolucionario (MNR). It began to agitate forcefully against the big mining companies and for expropriation. A strike in 1942, at Patiño's Catavi mine, was suppressed with gunfire which killed 19.

In 1951, the MNR ran one of its founders and longtime leaders, Victor Paz Estenssoro, for the presidency and, with the support of other leftist parties, he won. One of the platforms of the party and of the candidate had been to expropriate the big mines, and the nationalization took place in October of 1952. Aramayo, Patiño and Hochschild lost all of their mines and ancillary properties. The preamble to the expropriation decree blamed 'the big three' for just about every conceivable evil, existing or imagined. History has, however, by now given its verdict. The dismal state of the Bolivian mining industry in the 1980s, the run-down condition of its installations, the lack of exploration for new mines and the misery of its workers can not now any more be blamed on the 'tin kings'. But would the mines have been in better shape, had they continued in private hands?

Some 20 million dollars were eventually paid in 'indemnities' to the former owners.

The Mina Matilde story

The zinc/lead (but mostly zinc) deposit called Mina Matilde has been known at least since the early 1920s. It is located right on the shore of Lake Titicaca. The Mauricio Hochschild organization held the concession for some 30 years but operated the mine only on a small scale for high grade lead production. Over the years they invested several million dollars in exploration and mine development and would have put the mine into full operation, had the Bolivian government not expropriated the deposit in 1952.

It became the property of the Bolivian state. No further work was done until the mid-1960s. In 1966, two US firms, US Steel and the Philipp Brothers Division of Engelhard Minerals & Chemicals Corporation, signed a 20 year lease contract with the Bolivian government, then headed by President Ovando, to put in the necessary installations and infrastructure, and to produce.

Mina Matilde Corporation (MMC) was owned 50/50 by the two US parents, who supplied the basic equity. This was insured by OPIC, a US government insurance company. MMC next obtained a $13 million loan from Citibank and Bank of America, each supplying half. These loans were also insured by OPIC. The funds were used for mine development.

The deposit was put into production several months ahead of schedule and concentrates began to be shipped in the first months of 1970.

The location of the mine on the shore of Lake Titicaca, and the existing railroad from the Peruvian lake port of Puno to Matarani on the Pacific Ocean made the routeing of the concentrates across Lake Titicaca and down to Matarani a logical choice. As the existing transport facilities on the lake were insufficient, a new train ferry had to be built.

Production began in 1969 and quickly reached 8000 tonnes of zinc and 600 tonnes of lead concentrates a month. Mina Matilde Corporation was to operate the mine for 20 years in accordance with the contract signed in 1966. But after only two years, a new government in Bolivia, this one headed by General J J Torres and other leftists, decided to re-expropriate. But not the mine, because ownership of it had always remained with the Bolivian nation ever since the original 1952 expropriation. This time, the government was after the installations the partners had set up.

After MMC was nationalized, US Steel, the Philipp Brothers Division of Engelhard Minerals & Chemicals Corporation, and the banks made a claim against OPIC. OPIC recognized the claim but there was a waiting period of several months. OPIC used this to put pressure on the Bolivians (through the US embassy in La Paz) for a friendly settlement.

This was eventually achieved. Philipp Brothers got a long term Mina Matilde ore contract at competitive terms and were appointed retention agents. Out of the proceeds from concentrate shipments, Philipp Brothers repaid the banks, US Steel and itself, and in full. OPIC did not have to pay a single cent.

Although everybody got their money back, Philipp Brothers and US Steel did not, of course, achieve their original objective, i.e. to secure a captive production.

Production at the Matilde mine decreased steadily and eventually stopped completely.[25]

A summary, 1950–90

Nothing will illustrate better how the Bolivian mining industry fared after the 1952 revolution than a cold statistical appraisal of what was produced. For the principal metals, the figures are given in Table 7.1.[26,27] It is quite obvious that not much has happened. Tin production, always the mainstay of the country, has declined. Silver has risen somewhat but its present continues a paltry

Table 7.1. Mining production figures, 1945–91 (tonnes)

Year	Tin	Silver	Lead	Zinc
1945	43 200	208	9 500	21 000
1948	37 000	235	25 600	21 100
1953	35 400	190	23 800	24 000
1958	18 000	188	22 800	14 200
1963	23 100	151	20 200	4 600
1968	29 600	161	22 300	11 800
1973	18 400	164	20 100	48 900
1978	30 900	195	18 000	59 600
1983	25 300	187	7 700	47 100
1988	10 500	231	12 500	57 000
1989	15 800	295	15 700	74 800
1990	17 300	310.5	19 550	85 930
1991	16 800	337	20 800	129 800

shadow of its glorious past. Lead production has declined, but in zinc substantial growth has been achieved in the last 10 years.

The country has made some progress with smelting and refining operations, with the ENAF tin and antimony smelter, a few other small antimony smelters (Vinto and Palala), Telamayu's bismuth smelting and refining plant, as well as the Hormet tin fuming plant and lead smelter, in addition to the old Fundestaño de Oruro tin smelter. There is also the Karachipampa lead/silver smelter owned by Comibol and ENAF. Some of these works are only operating on and off; a few have been out of production for many years.

Only recently, from 1989, did Bolivia make serious efforts to get its usually volatile currency and inflation under control, and maybe a new effort at democracy and free enterprise policies under the current President Jaime Paz Zamora will get the country and the mining industry back on to a steady course so that new investments are made and production will again increase.

And gold?

Among the Latin American countries, Bolivia occupies the sixth position as a gold producer. It should strike us as odd that this prime mining country, which is the second largest tin producer on the continent and, behind Peru and Chile, South America's third largest silver producer, should have so little gold.

Production in 1990 reached 10.4 tonnes, up substantially from only 2 tonnes in 1980. Perhaps it is that there is more gold produced, but it is not reported?

Notes

1. The question of much the pre-Columbian people knew of metals and metallurgy remains open. Dr Georg Petersen, in his book *Mineria y Metalurgia en el Antiguo Peru*, lists the Quechua language words for some 40 metallic and mineral substances, implying, if not claiming, that they were all known by the time of the conquest.
2. History occasionally allows interesting speculations. We can wonder what would have happened had there been large scale German colonization of South America beginning about 1550, or an Almagro government instead of the civil wars that ravaged these countries for many years.
3. It must have been rosiclair ore.
4. Cañete, *Historia Fisica y Politica de la Provincia de Potosí*, page 51.
5. Costa Villavicencio, *Historia Cronologica del Peru, 1500–1541*, page 53.
6. At the surface, the rich ore was mostly rosiclair, oxide ores that were just melted down. Once the Spaniards were beyond the surface, into the mountain, the sulphide ores gave them trouble.
7. Ballantine Cobb, page 31. Ballantine Cobb's book, *Potosí y Huancavelica*, is a remarkable example of historical analysis and writing at their very best. Potosí and the extraction of its silver are inconceivable without the mercury from Huancavelica – and Ballantine Cobb made this the subject of her book. Thus, mining in Peru and Bolivia was reviewed together, at least for the period 1545–1640.
8. 'Assay' undoubtedly meant that Garcés treated the ore and obtained mercury.
9. All of the following calculations should be taken with a grain of salt and the resulting totals should impress the reader that what was produced was simply fabulous.
10. This comes from Ballantine Cobb who cites Pedro Cieza de Leon.
11. Ballantine Cobb, page 104.
12. Capriles Villazon, pages 56–7. Orlando Capriles Villazon's *Historia de la Mineria Boliviana* is a remarkable book, not the least because it is about the only country-wide review of an industry, or better, this industry, in Latin America. While many a historian reviews mining as part of a chapter, or part of several chapters, in different 'Histories' of several countries, Capriles Villazon sensed, and correctly so, that mining played a very important role in the history of Bolivia, and deserved thus a separate analysis, meaning a separate book.
13. Ibid, page 65.
14. Ibid.
15. Ibid. Luis Peñalosa actually used the calculations of Torrente.
16. Cañete, page 51.
17. Capriles Villazon, page 62.
18. Ibid, page 60. This is based on the premise that one reale had a weight of 3.43 grams, and a peso (of 8 reales, also called *comum de plata*) weighed about 27.44 grams.
19. Ibid, page 61.

20. The Aramayo story comes essentially from Alfonso Crespo, *Los Aramayos de Chichas: Tres Generaciones de Mineros Bolivianos*.
21. The Patiño story can be found in any book on mining in Bolivia. The best account, and the one to which I am heavily indebted in this section, is by Charles F Geddes, *Patiño, the Tin King*.
22. One of Hochschild's former employees wrote to tell me that Don (as he used to be called) Mauricio Hochschild personally told him in about 1940 that Huanchaca had been, 'back then', really a working mine. He had tried to get the management of it, but as he owned only some 15% of the shares, the majority owners refused to hand the management over to him. But Hochschild had found out that the exploitation rights to the mine did, by law, not cover the dumps (the leftover, discarded ores of several centuries of work by previous owners). Hochschild filed claims to the dumps and then refused passage to the miners who wanted to enter the mine, claiming they would trespass. The majority owners had to negotiate and did eventually turn the management over to Mauricio Hochschild. Even if apocryphal, the story could be true. It illustrates the rough and tumble atmosphere in the Bolivian mining scene in the 1920s and 1930s.
23. I worked for the Hochschild group in 1953, and again in 1956–9, in New York and Lima, and subsequently maintained contact with many former colleagues. This enabled me to gather data on the origins and growth of the Hochschild group.
24. A somewhat personalized light is thrown on some of the people and events described in this chapter by a letter I received from Rolf Kantorowicz who, for many years, was associated with different ore-buying firms in Bolivia and Peru. He knew many of the actors personally. His comments go beyond mere personal reminiscences and are reproduced here, with his permission:

> When the tin price declined in the second half of the 1930s, the tin pool saw the need to support the price and thus fixed limits, or export quotas, for each member country. In Bolivia, these quotas were called *cupos* and were issued proportional to the production of each mine.
>
> Small and medium miners were entitled to export 10–15% of the overall Bolivian quota, the Aramayo group 10%, Hochschild 25%, and the remainder of about 50% went to Patiño.
>
> None of the groups felt it had enough quota to continue orderly production; Hochschild especially tried by all means to increase his share. But this could only have been possible at the expense of the largest quota holder, Patiño, who did not agree to cede anything.
>
> Hochschild and Patiño did not get along with each other and the dispute soon led almost to 'war'. But the advantages were all with Patiño whose production was larger, who controlled smelters and had by far the greater financial resources available. As the Patiño group was also shareholder in many of Hochschild's companies with, in some cases, even a seat on the board (in Cia Minera Oruro), Patiño managed to avoid seeing his quota reduced.
>
> It was only in the *rescate* business that Hochschild dominated while Patiño did not participate in this business at all.
>
> Early in 1937, Banco Minero de Bolivia began with its own *rescate* activities, but without much success as the bank had neither qualified personnel nor sufficient funds. As a consequence, the bank became private, in 1938. Patiño acquired 60% of the shares, the Bolivian government stayed with the other 40%. New managers were hired, mostly foreigners, and, with the backing of Patiño's power, the bank now made headway.

In June of 1939, the government 'nationalized' the ore and mineral trade and export; only Banco Minero could henceforth export. But in practice, it still did not work as the bank still did not have the capital nor the infrastructure nor the international connections necessary to successfully trade in ores. Thus, a few months later, the law was changed and the large groups were back in business, more or less as before.

Hochschild saw the ore business endangered and tried to establish a consensus among the producers to establish that they would prefer to export through the big firms. But this was 'high treason' in the Bolivia of those days, as the expropriation law of 1939 had stipulated, and with which the trade had been monopolized in the hands of the bank.

As a consequence, Mauricio Hochschild himself was arrested together with Gerardo Goldberg and Adolfo Blum of his group and even Arturo Gruenebaum of Philipp Brothers, as co-conspirators.

The cabinet of the country, under the direction of President of the Republic Colonel German Busch condemned Hochschild, Goldberg, Blum and Gruenebaum to death. Six ministers supposedly voted for the death penalty, six against, and Busch tipped the scales. Busch now ordered the chief of police to report but the man had waited endless hours and had finally left to have lunch – whereupon the execution was postponed for the next day by which time the US and Argentine embassies intervened (Hochschild was by then an Argentine citizen). Also private people intervened and Busch was eventually prevailed upon to suspend the executions. As he had also a priori ruled out a monetary fine, the four were eventually set free.

But the export monopoly of Banco Minero continued and Patiño soon ended his brief entry into the *rescate* business. And also Hochschild's *rescate* activities soon ended – in Bolivia. In Peru, Brazil, and Chile they continued for several decades.

25. Henry Rothschild, of Hartsdale, New York, cleared up many questions I had about the Mina Matilde story. Mr Rothschild was one of the senior executives of Philipp Brothers involved in the story.
26. These production statistics are once more from Metallgesellschaft and their detailed annual book *Metallstatistiken*, except for those for 1990 which are from the Ministry of Mines in La Paz.

    A word of caution is in order. The Ministry of Mines only has export statistics; reliable production statistics seem to be missing. The data I got from Bolivia pertaining to years prior to 1990 are very different from those Metallgesellschaft published. I have used the statistics of Metallgesellschaft, wherever they were available.

    Regarding gold, no reliable data exist apparently. From Bolivia I was told that the country produced something like 5–15 tonnes per year during the last 10 years. The official statistics of exports seem to show something like 5 tonnes per year. Smuggling is as important in Bolivia as elsewhere.
27. The decline in production of tin has been severe. Some experts say the country is simply 'worked out', that there is no more tin, and that the future will bring still more decline. But there is more ore, only it is of low grade. And at prevailing low prices, it is not economical to produce.

CHAPTER 8

In praise of horses, llamas, camels – but mostly mules

The pre-Columbian natives had no beasts of burden other than the llamas of the Peruvian and Bolivian highlands. These may also have been used in Equador and southern Colombia, as well as in northern Chile and northern Argentina.

The Spaniards brought horses soon after Columbus' first trip and they also brought other domesticated animals, such as hogs, cats and dogs as well as cattle. Many of the cattle were allowed to roam freely, and reproduced spectacularly, to such an extent that by 1580 cattle hides were the principal export article of Hispaniola.[1]

And the Spaniards also brought donkeys (*burros*), which also reproduced spectacularly, especially by crossbreeding with horses to get mules. Horse reproduction on Hispaniola was at such a high rate – and apparently well known throughout the Caribbean – that as early as 1530 a caravel put into Santo Domingo to try to get horses for Pizarro's planned trip to Peru.

Horses were one of the weapons – if not the principal weapon – the Spaniards used during the conquest. Whereas Cortés had still used only a few (16 horses for 508 men), Pizarro had 62 horses for 172 men, which provided a formidable force. The Spanish horsemen had been the fear of Europe, with their light mail and 7 foot lances.[2] They must have scared the living daylights out of the natives. And the Spaniards played up to the magic fears of the natives – even to the extent of claiming, at least for a short while at the beginning, that the horses needed gold as fodder. And the Spaniards also took great care to avoid the horses being wounded; and, if they were, to avoid the natives seeing a dead horse.

Spirited charges by horsemen settled, at least for a while, many a battle of the almost interminable civil wars between different factions of the conquistadors.

It was from about 1530 on, in Mexico, and from about 1550 in Peru and Bolivia, that horses, and especially donkeys and mules, came into their own as beasts of burden. The entire system of commerce between the interior and the ports was based on the availability of mule trains. They took the gold and silver ores to the smelting points, where the metal was made in small furnaces. And they then took the bars to the ports. They carried supplies, principally mercury, for amalgamation, tools, and all and everything the mines and towns in the interior needed.

Traffic between the port of Callao and Lima was unthinkable without mule trains, let alone traffic between the mines and Mexico City, and on to Veracruz.

Even the treasures of the Far East, which the galleons brought to Acapulco, had to travel from there by mule train across Mexico to Mexico City, and then to Veracruz, for onward transportation to Europe.

Camels were tried in several countries, for a short while. Proof exists that they were introduced into Cuba in 1550 and into Peru in 1565, but their contribution to the local transportation was minor and their reproduction difficult. They simply died out.

The scope of colonial mining depended strictly on mule – or horse – transportation capabilities. Llamas never carried more than 50 kilograms, and only the males could be used. Their kin, alpacas and vicuñas, were not usable. But mules were prime carriers. They could carry as much as 150 kilograms on more or less level ground, although not more than 100 kilograms on mountain tracks, and could go as much as 30 or 40 kilometres without much pasture or water.[3]

The system was that mule trains would consist of units of 20 animals each under the direction of two drivers. Only half of the animals would carry loads; the others would string along as spares. No mule train owner or driver could, obviously, risk losing an animal on the way and have no other to pick up the load. What would he do – abandon the load? The trails in Peru, and presumably also in other colonial countries, were signposted by bleaching mule and donkey bones.[4]

There were, of course, no roads in the Americas. The natives had not known the wheel, but even after it was introduced by the Spaniards, it served little purpose. Almost everywhere the topography was accidented – like crumpled paper as Cortés (or Pizarro) had explained to the emperor – and wheels are only of use on more or less level ground. In Europe, by comparison, a horse-drawn cart (with four to six horses) could transport 1.5 tonnes; on fully level ground the load could go as high as 3–4 tonnes.

To Lima and Potosí were imported something like 100 000 mules per year, via the overland route from Tucumán and Córdoba, in the mid-eighteenth century. In about 1770, Mexico imported something like 70 000 mules a year to Veracruz and 75 000 to Acapulco.[5]

The beasts were needed for all purposes. We should, for instance, think of the patio process, widely in use by the year 1600 in the Mexican, Bolivian and Peruvian silver mines. Once the mixture of finely ground ore, plus salt, copper sulphate and other ingredients was ready, it had to be well 'kneaded', i.e. beaten or trampled – in some cases for weeks. This trampling was at first done by human – native – feet, but soon by herds of horses and even more by mules, until sufficiently 'leavened'. We can hardly imagine what this interminable exposure to mercury did to the hooves of the animals – not to speak of what it did to the feet of the natives in the early years.

The silver of Potosí and other Peruvian and Bolivian mines went in the beginning to Lima, but later mostly to Arica and then, by vessel, to Panama. From there it had to cross the isthmus to the other side, at first to Nombre de Dios and later, after that town had been abandoned because of the fevers, to Portobelo which must have been healthier and could apparently also be better defended against attacks from pirates, although they did plunder it repeatedly.

From Panama, where the vessels discharged their cargo from Peru and Bolivia, the route went up along what was called the Camino Real, the 'royal highway'. It was an almost straight road, running for some 30 kilometres, to a town (or place) called Cruces. The road was built to suit animal traffic, with suitable ascents and descents for donkeys, mules and horses – and, of course, for human guides and porters on foot.

'Road' is probably a misnomer. It was not built for any kind of vehicle; it was probably no more than 4 or 5 feet wide, and had many steps – a road such as can still be seen in a few places in Peru where parts of the old Inca highway system have survived. This was probably not much different from the Spanish Camino Real in Panama.

By 1850 – and we have reports of crossings from the fortyniners, the prospectors who rushed to the newly discovered gold placers in California in 1849 – the old stone base of the Camino Real had disappeared and the trail was little more than a narrow jungle path only safely negotiable by sure-footed mules.

After Cruces, the route became a river, the Chagres, for some 60 kilometres – by barge or canoe – all the way to Portobelo.

After 1855, when the railroad across the isthmus was completed, everything changed. The crossing now became quick and easy.[6]

In 1857, the Americans tried to reintroduce camels, for the arid areas of the south-west and west. The animals were imported from Egypt and Turkey and were then taken, probably from New Orleans, across Texas and Arizona to California. With the camels came a Syrian, Hadji Ali. As the name was unpronounceable to American tongues, he quickly became 'Hi Jolli'. A monument to him can still be seen in Quartzite, California.

Another vignette, this one about mules and oxen is interesting. Mule trains crossing the American midwest could travel 40 miles a day, on flat country. Oxen did only 15 miles, but were generally used because they were more reliable and pulled heavier loads. And oxen ate food that mules refused. Still, even the oxen occasionally ran into trouble. Clover, blue grass, herds grass and buffalo grass could be eaten by sheep, mules, horses and bison. These animals have flexible snouts. But oxen have stiff snouts and can only graze to 1½ inches above ground. And buffalo grass rarely reaches that height.

Beasts of burden – principally mules, but also horses and, on a minor scale, llamas and camels – were essential for the development of Latin America. In North America, oxen and horses pulled wagons across the plains and into the mountains, but in the mountains themselves, it was once more mules and horses.

Mining was unthinkable without such beasts of burden – to bring supplies in and, mainly, to take the production out.

Notes

1. Cassá, *Historia Social e Economica de la Republica Dominicana*, page 136.
2. Peterson, *The Funnel of Gold*, page 2.
3. Descola, *Daily Life in Colonial Peru, 1710–1820*, page 223.
4. Ibid.
5. Semo, *Historia del Capitalismo en Mexico*, page 71.
6. Jackson, *Gold Dust*, pages 76–9.

CHAPTER 9

Chile

The conquistadors did not find much

As we saw in Chapter 7, Diego de Almagro, one of Pizarro's partners in the conquest of Peru, obtained from the emperor the governorship of 'everything south of Chincha, and 200 leagues inland', the same area that had previously been given to the Fuggers.

Almagro suspected there was, or hoped for, considerable wealth to the south, and he set forth from Cuzco on 3 July 1535 with 570 horsemen and many footsoldiers, as well as 12 000 native troops. Almagro was then already 55 years old, and his empire, or country to be conquered, was to be called Nueva Toledo.

They travelled inland, past where Puno is today, along the south-western shore of Lake Titicaca, along the Desaguadero river, and hence through what would later on be La Paz, across the Cordillera de Chichas to Tupiza, Jujuy, Chicoana and Santa Maria, to then turn straight west, across the Andes to Copiapó, the Chilean port. Here they stayed a while and, lo and behold, a vessel sent south from Callao brought them fresh supplies.

They continued on to Huasco and Coquimbo and then all the way down to Valparaiso.

They encountered many ferocious natives, also extremely cold temperatures in the Andes – but of gold and silver they found very little. Near today's Santiago de Chile they decided to turn back. Straight north they walked; they crossed the Atacama desert and continued via Arica and Tacna to Arequipa, in southern Peru. The trip had taken almost two years. The result was zero.

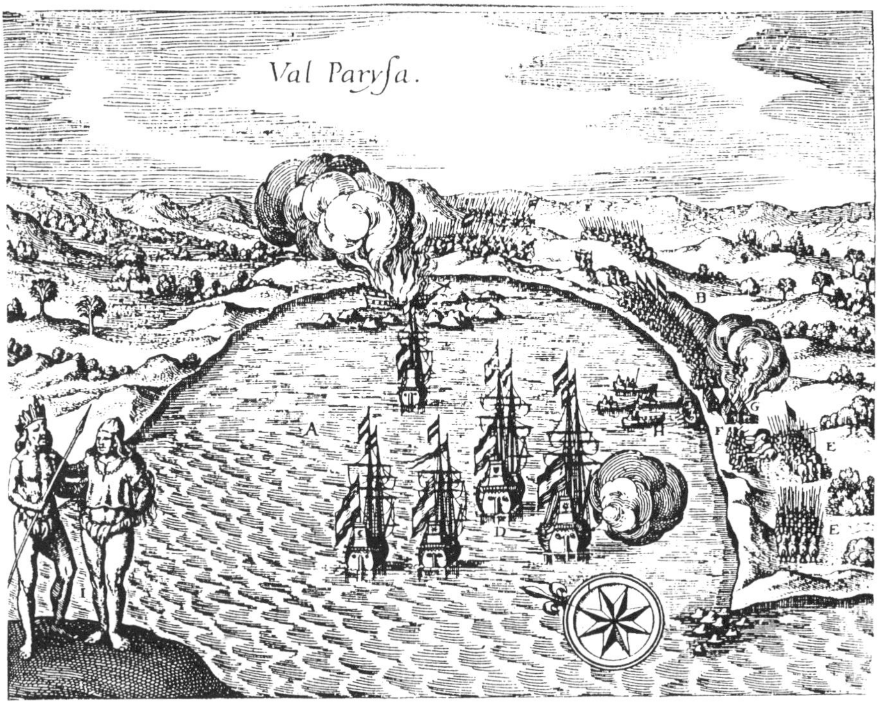

A view of the port of Valparaiso in about 1615. Source: Johan Ludwig Gottfriedt, *Newe Welt und Americanische Historien*, first published in 1631 by Matthaeus Merian dA and republished in 1980 as Stuttgarter Faksimile Edition im Fackelverlag, Stuttgart, Germany.

But not quite. There are stories that they passed near the Chuquicamata deposit, then already worked by the natives – although on a minute scale – and that they had their horses' hooves shoed with copper. And Almagro had passed through and had taken possession of much of what is today Bolivia and Chile, passing near several of the large silver and copper deposits that were subsequently to become important.

It has also been said that here, in northern Chile, the natives knew how to smelt copper (or maybe just melt native copper?), how to harden it, and even how to make bronzes (presumably with tin from the mountains). Recent finds seem to prove that the natives used copper some 2000 years before the conquest, or in around 500 BC.[1]

A colonial backwater

We should not doubt that copper was produced in Chile throughout the entire colonial period, but it was not much.

Corporación de Cobre, Codelco, published in 1975 an excellent summary of the origins, the history and the (then) current situation of the industry in the country. According to Codelco's tabulations, about 81 550 tonnes of copper metal were produced during the 210 years of colonial rule 1601–1810, for which some records could be found. Minute quantities were also produced during the sixteenth century.

But Chile was a backwater during the colonial period.

To begin with, it took 330 years, until 1870, before the fierce southern tribes beyond the Bio Bio river were dominated. Much effort was spent subjugating them. Also, part of what is Chile today, between Arica and the Loa river, was Peruvian until the Nitrate War[2] of 1879–83, and Bolivia held the area south of the Loa river all the way down to Taltal.

During much of the colonial period, Chile was a rich supplier of wine, olives and other foodstuffs, and also mules. And some of the southern border towns and garrisons were used as prisons for political, criminal and even religious and moral prisoners. But there was some mining.

Copper was used as ballast in the ships that went south around Cape Horn to Spain, or north to Peru. In 1651, the Peruvian Viceroy Conde de Montes Claros asked for copper from La Serena to make cannons to fight the pirates.[3] The viceroy wanted 30 tonnes; it took almost 10 years to complete the order.[4]

Most of the colonial copper came from Coquimbo and Valparaiso. The Coquimbo (La Serena) copper was considered 'sweet', of great purity, hence malleable, and the Valparaiso (Aconcagua) copper was impure, hard and hence most suitable for bells and cannons. But, in general, sulphide ore could not be processed: it was only surface oxides that could be smelted with charcoal.

In 1707, gold was discovered near Copiapó, and a great rush began. But it lasted only a short time and soon most mining activities ceased. It came to a point where the crown, in 1785, issued a new mining code to stimulate private investment.[5]

Still, it did not amount to much, and Chile cannot be considered a prime mining country during the colonial period.

It was only after the independence wars that production really picked up. From about 1500 tonnes a year in around 1810, Chilean production went to 6500 tonnes of copper a year in about 1830 and to about 20 000 tonnes a year 25 years later. By then Chile had become the number one producer of copper in the world, a position it held for about 30 years before it ceded the position to the United States. Ever since, Chile has generally remained the number two copper producer, until 1982 when it again became number one.

Table 9.1. Copper production, 1851–80 (tonnes)

Period	Production	Average per year
1851–60	220 000	22 000
1861–70	401 000	40 000
1871–80	457 000	45 000

The Chañarcillo silver rush, 1832–41

While Mexico, Peru and Bolivia, as well as Colombia and Venezuela, went through more or less prolonged periods of political instability – which also brought, as a consequence, economic stagnation that ended in Mexico only with the arrival of Porfirio Diaz in 1876 and in Peru only with the arrival of Ramón Castilla in 1845 and in Bolivia perhaps never – Chile benefited from extended periods of good and stable government. Foreign investors were invited in, got favourable conditions and invested. They also brought mining experts and new mining techniques – especially the British. Almost the entire northern part of the country – then only extending to Taltal – was searched and new mines discovered. A French chemist, C S Lambert, became famous when he set up, in 1831 in Guayacan, the first prototype of a reverbatory furnace where sulphide copper ores could finally be smelted. In 1841, he also put up the first copper and brass rolling mill at La Serena.

Among the new mines discovered were, in 1832, silver deposits in Chañarcillo. During the following decade, these mines produced about 1 million ounces (about 30 tonnes) of silver per year, and from 1850, about 200 tonnes a year.[6] These finds, and other mines subsequently discovered, gave Chile a silver production of roughly 8000 tonnes during the nineteenth century. This wealth, in addition to probably some 10 million ounces (about 300 tonnes) of gold,[7] helped finance two wars and allowed the country to pay off its foreign debt.

Copper is big business

Table 9.1 shows, at a glance, how Chile's copper production grew. The best year was 1876, with 52 303 tonnes.[8] The next big jump was only to come in 1910.

Chile was now a world power in copper. While it was the number one producer, it accounted for 35–45% of total world production.

But from about 1880, production began to decline again, partly because the mines were becoming deeper and more difficult to operate, and also because the grades declined. And, finally, because more money could be made with nitrates.

The Nitrate War

Large quantities of sodium nitrate, a salt used as a fertilizer and for explosives, had been discovered in the desert coastal region of (then) southern Peru and western Bolivia, roughly between today's cities of Arica and Antofagasta, in the so-called Atacama desert. As it practically never rains there, the nitrate stays on the surface.

Two gold prospectors supposedly once, a long time ago, lit a fire in this desert, and the ground lit up with blue flames.

Today, synthetic nitrate is made from nitrogen in the air and has been since 1880. But in the period 1860–80, nitrate from South America was supreme. As Bolivia paid little attention to its nitrate deposits in what was then its province of Antofagasta, Peru was too preoccupied with its internal problems to do so, and both countries were perhaps too sparsely populated to be able to occupy and settle these coastal provinces adequately, it was Chilean capitalists and Chilean workers who settled there and worked the mines.

Eventually, in about 1877, first Peru expropriated the mines and then Bolivia raised taxes on nitrate production from mines in its province – and the Chileans went to war.

The Nitrate War lasted from 1879 to 1883; Chile soundly trounced both Bolivia and Peru and despoiled them of the entire coastal region to just north of the town of Arica. Chile not only obtained the nitrate deposits but also – though these were not yet known then – the valuable copper deposits of Chuquicamata, Exotica and Mantos Blancos – plus whatever else is still to be discovered.

Decline, 1880–1910

The victory in the war in 1883 lifted Chile's spirits up. But it was the spoils from the nitrate mines that brought prosperity to the country, and dimished interest in other mining. If investment in the north, in nitrate mining, and imports to spend the easily earned foreign currency, brought excellent returns, why bother about copper? Mining was thus neglected.

By the year 1900, 748 mines were being worked, yet only 50 of them were in any way mechanized.[9] Production declined accordingly, to 25 000 tonnes of copper in 1889, and to 21 000 tonnes in 1897. Thus, by 1897, Chile was also down to 5.2% of total world copper output.[10] And it was down to sixth place in the world. But changes were about to come.

Major foreign investments, 1904–27

Good ideas often come when the time for them is ripe. Thus, an American mining engineer, Daniel Jacklin, just about then sold the Guggenheim brothers the idea that 'economics of size' would make sense in copper mining. Thus, 'if one has a bigger plant, putting through more ore, one could treat lower grade ore and still make money,' he said.

And the Guggenheims put the idea to the test, at Bingham Canyon in Utah in 1906/7, with first a 2000 tonnes/day, and then a 6000 tonnes/day plant. These were large quantities of ore. With time, Bingham Canyon grew to 15 000, 30 000 and by now 110 000 tonnes of copper ore per day. Bingham Canyon was the first open pit, large low grade porphyric copper deposit to be worked.

Soon another American mining engineer, Willian Braden, went to Chile and looked at deposits called El Teniente (near Santiago de Chile) and Potrerillos (near Chañaral); others looked at Chuquicamata (near Antofagasta) and similar large scale low grade deposits.

Soon companies were being founded, El Teniente in 1904, Chuquicamata in 1912, and Potrerillos in 1920.

The deposits were amazing – and remain so. El Teniente was originally set up with a 250 tonnes/day plant, and is still going strong today with a 66 000 tonnes/day plant.

Chileans claim that it was at El Teniente that selective flotation was tried out for the first time, and successfully, in 1916. With this technique, copper recoveries improved from 60%, with the traditional gravitational method, to 85–90%.[11]

Chuquicamata, run by the Guggenheims, went into production in 1915, at once on a large scale. Potrerillos went into production in 1920, and went out of production in 1959, to be replaced by El Salvador.

Two world wars

The Guggenheims who, as in Mexico, had in Chile stolen a march on their competitors by setting up El Teniente and Chuquicamata, sold both: the first, in 1915, to Kennecott, and the second, in 1923, to Anaconda.

By the 1920s, Chile was back in the running as a major copper producer, with a yearly average of 200 000 tonnes which rose to 270 000 tonnes in the 1930s and 435 000 tonnes annual average in the 1940s, when it reached a high of 18% of world production and second place behind the United States. After World War II, Chile settled into a position of supplying 10–15% of the world's copper needs.

'Chileanization', 1970–3

Chile, like other South American countries, also caught the bugs of nationalization, xenophobia, 'what's ours is ours', 'gringos out' and so forth, in the

Table 9.2. Ownership of the Chilean copper industry

Name of mine	Operator	Owner
El Teniente	Braden Copper Co	Kennecott, a US company
Chuquicamata	Chile Exploration Co	Anaconda, a US company
Potrerillos/El Salvador (1960)	Andes Mining Co	Anaconda, a US company

years 1955–75. It almost seems that there is a clear pattern, a cycle, that every country must sooner or later go through. The cycle looks like this:

1. At first, nobody is interested in mining.
2. National, or domestic capital is either unavailable or the country is unwilling to risk it.
3. Foreign capital is invited.
4. Foreign capital invests.
5. Foreign capital succeeds.
6. National interests become jealous.
7. 'Let's expropriate.'
8. Nationalization.
9. Euphoria.
10. Hangover.
11. Stagnation.

And then the cycle begins all over again.

Eduardo Frei, of the Christian Democrats, became president of Chile in 1964. The ownership of the Chilean copper industry looked then as in Table 9.2. The three companies mentioned for many years produced about 90% of all Chilean copper. By about 1965, their share was down to about 80%. The rest came from Disputada de las Condes (from 1916), owned by Peñarroya, Mantos Blancos (from 1957), owned by the Mauricio Hochschild group, as well as from a few smaller operations.

It was Frei and his party who introduced the concept of 'chileanization'. They wanted gradually to become first partners of the foreigners and then take over completely. It took seven years, and much back and forth, and eventually much coercion, but by the end of 1971, under President Salvador Allende, the copper companies were foreign no more.

To calculate the price to be paid for the mines, Chile took the book value of the companies, less revaluations made since 1965, less certain amounts for 'maintenance, repairs and parts replacements not carried out by the owners' and, furthermore, less 'excessive profits', namely those profits the companies had had in certain years and which had been in excess of 10% of the book value – 10% being considered the limit of reason.

Table 9.3. Copper production, 1965–91

Year	Production (tonnes)
1965	584 000
1968	657 000
1970	692 000
1971	708 000
1972	717 000
1973	735 000
1974	902 000
1975	828 000
1976	1 005 000
1977	1 056 000
1978	1 034 000
1979	1 063 000
1980	1 067 000
1981	1 081 000
1982	1 242 000
1983	1 256 000
1984	1 291 000
1985	1 356 000
1986	1 400 000
1987	1 418 000
1988	1 451 000
1989	1 609 000
1990	1 599 000
1991	1 814 000

The calculation led to the absurd extreme of El Teniente owing US$ 310.5 million to Chile – meaning Chile paid nothing for the expropriation. Also Chuquicamata was calculated to owe $76.5 million.[12]

The political and social upheaval and the economic decline which followed affected the copper industry first and foremost. Productions was soon down to 60% of the pre-nationalization level.

More spectacular growth

But Allende did not last. After decline and stagnation, Chile entered, under the Pinochet government, another period of reassessment and *laisser-faire*. And though ownership of the big mines was not returned to the big companies, foreign capital was made welcome once more and did flow back into the country.

When looking dispassionately at the statistical picture, the confusion of the Allende years passed relatively quietly and without major damage to the growth of the copper industry. Table 9.3 shows the production figures for 1965 to 1991.[13]

And copper is not all Chile produces. The country is also the world's number one producer of rhenium with, recently, some 6–7 tonnes per year. In molybdenum, Chile is number two, with 15 000–20 000 tonnes per year. Both are by-products of copper production.

The quantities of lead and zinc are modest – about 1000 tonnes per year for lead and about 30 000 tonnes of zinc.

Substantial quantities of silver are being produced – also mainly as a by-product of copper production. The 1991 production reached 673 tonnes. Gold production rose steadily to 31.6 tonnes in 1990, but then declined again to 21 tonnes in 1991.

Chile's future, or rather, the future of Chile's mining industry, looks bright. In addition to the already existing operations which promise continued large production, a number of other copper *deposits* are known and will probably in time be put into production. These are Mocha, Cerro Colorado, Quebrada Blanca, El Abra, Pampa Norte, La Escondida and Andacolla. It is ironic that all, with the exception of Andacolla, lie in the northern provinces, those that used to belong to Peru and Bolivia, until the Nitrate War.

Notes

1. Corporacion del Cobre, *El Cobre Chileno*, page 5.
2. Also called 'War of the Pacific'.
3. Corp. del Cobre, page 15.
4. Ibid, page 17.
5. Ibid.
6. Ibid, page 23.
7. Ibid, page 25.
8. Ibid, page 23.
9. Ibid, page 27.
10. Ibid, page 29.
11. Ibid, page 31.
12. Ibid, page 51.
13. These details come once more from the Metallgesellschaft statistics.

CHAPTER 10

In praise of railroads

As long as mining was based on human or animal transportation – with mules, horses, camels and even llamas – progress was slow because there was just so much a mule or horse could carry. There were limits on quantities and limits in terms of cost. And these limits restricted mining to metals of high value, gold and silver, which indeed were the ones the Spaniards exploited between 1492 and about 1850. (The case of mining cinnabar ore, to get mercury, at Huancavelica in Peru, and also in Mexico, was discussed, but this was something special, required precisely for the production of silver at nearby mines.) The presence of other metals was known – lead, copper, antimony, iron, and no doubt also zinc, bismuth, and many more – but mining them, converting ores into metals, and then transporting them down from the Sierra – where the mines were generally located – to the coast and then on to Europe made no economic sense. We know of reports that copper and lead slabs were used as ballast in ships, but these cases were few and meant little in terms of mining.

It took the Industrial Revolution, the industries and machinery and equipment it created, and the then rapidly growing demand for more and more raw materials of more and more types for more and more applications to set in motion demand for and the production of what are generally called base metals.

The Industrial Revolution commenced in England, about 1770. The first breakthrough was in textile production, with new spinning methods and steam powered weaving looms for, at first, cotton, but soon also for wool. The invention of an efficient steam engine was fundamental for industrial growth.

It allowed setting up factories; it allowed more efficient coal mining; it allowed more efficient water pumps; and – the next step – it allowed for movable machines once stationary ones were proven and successful. Locomotives, and railroads, were the result.

The age of railroads began in 1830 when George Stephenson's locomotive made its first run between Manchester and Liverpool.

Actually, railroads – or steam driven locomotives used for industrial purposes – had come into use already earlier, in 1804, in Wales. And the inventor, one Richard Trevithick, soon travelled to Peru – and the connections between steam engines, steam pumps, locomotives and their usage in mining were thus established very early.

Why Peru? Peru was known as the heartland of the Spanish possessions in the Americans. It had been the centre, and the origin, of the (supposed) Spanish mining wealth, explored and exploited for three hundred years. It mattered little that most, if not all of the glitter had already disappeared. What mattered was the reputation. It must be remembered that Peru, indeed all of the Spanish possessions in the Americas, had been off limits for all foreigners until the year 1810. Inaccessibility had increased the mystery: the wealth had to still be there, foreigners now only had to flock there to get it.

It is thus completely logical that inventors of equipment and machinery suitable for mining travelled to Peru to try to sell their inventions and to try to participate in a new mining boom.

Trevithick had sold several steam pumping engines to Peru. He had sold them on credit, and when he did not get paid, he picked himself up and travelled to Peru, in 1817 to look after his money.

His steam pumping engines had been destined to drain or dry the Cerro de Pasco mines – so that these could be worked again.

Instead of payment, Trevithick accepted a partnership in the project. But the engines did not work, the miners objected – fearing they would lose their jobs – and Peru submerged into its war of independence. Trevithick left in 1824, poorer and disappointed.

It is significant that Trevithick was in Peru – with his engines and ideas – and had it only been that the machines had worked and that Peru had not submerged into the war, pumps, locomotives and railroads could have come about in Peru almost as early as in England.

But with the unrest and the commotions in most of the countries of Latin America, it took until the early 1850s before railroads were built.

The first railroad in South America came about in Georgetown, Guyana, in 1849. The second was built in Peru, between Lima and Callao, in 1851 – as the first stretch of a line to lead up into the sierra, to connect the coast, and the port of Callao, with the Cerro de Pasco mines. It was 1893 before the stretch Lima–La Oroya was completed, and La Oroya–Cerro de Pasco was only finished in 1904. Railroad building in Peru was much delayed by the Nitrate War with Chile, 1879–1881.

Meanwhile, other railroads had been built including, also in 1851, the 75 kilometre stretch from Caldera to Copiapo, in Chile, and which was actually the first line to bridge, or cross, the Andes. And it was also the first to, in fact, be built to connect a mining centre with a port.

Other lines followed, like the Valparaiso–Santiago de Chile–Mendoza (Argentina) line, in 1910, the Antofagasta–Bolivia Railroad in 1892 to Oruro, the Arica–La Paz line in 1913, and the Southern Railway, in Peru, from Mollendo via Arequipa to Puno (1874) and Cuzco (1907).

Much of the Antofagasta–Bolivia Railroad owed its existence to mining, especially the stretches to Potosi and Oruro. The so-called Nitrate Railroads, those from Pisagua and Iquique going inland, as well as the Tocopilla line also owed their existence to mining, that of nitrates.

A strong case can, in fact, be made, that all of the railroad lines of western South America were built essentially to give logistics support to mining ventures which either existed or were expected to be set up once a railroad was in place.

If Richard Trevithick, a Cornishman, was the first to bring steam engines to Peru, a Yankee from Newburyport, Massachusetts was the next: William Wheelwright. As US Consul in Guayaquil, Ecuador, he urged people to acquire steam-driven vessels to move goods between South America and the US and Europe. The founding of the Pacific Steam Navigation Company in 1839, in Liverpool, is directly credited to his efforts. He also developed an early project for the Panama canal. But his most daring ideas concerned trans-Andean railroads, foremost among them one from a Chilean port to the River Plate.

But the most important – and most colourful, and a real entrepreneurial giant – of the foreigners involved with railroads in South America was another American, this one from New York: Henry (Enrique) Meiggs. His career was chequered, with many ups and downs.

Meiggs was one of the forty-niners who rushed to California in 1849 to participate in the gathering of easy wealth that was supposedly there for the picking. But Meiggs neither dug nor washed gold. Instead he went into lumber and made good money. He then moved into land speculation, and soon also into politics. He made still more money and became a patron of the arts.

But for all the assets he acquired, there were also liabilities: debts, and by 1854, Meiggs' debts were larger than his assets. Unfortunately, that was also the year when there was a crisis, even a crash. Some of Meiggs' debts were called, and he was 'short'. He next stole a booklet of promissory notes of the city of San Francisco – his brother had just been elected comptroller of the city – signed in advance by the mayor. Meiggs sold enough of them to get \$365 000, then gathered his family, boarded a vessel in the harbour, and skipped town. A warrant was issued and he was pursued. It took almost twenty years before he made good, and it took until 1977 before the

San Francisco Court of Historical Review and Appeals cleared his name.

Back in 1854, he sailed via Tahiti and other Pacific Islands to Chile, to begin there a second career, as *the* railroad promoter and builder of South America.

While US marshals pursued him in Chile with extradition papers, and he barely managed to dodge them, he bid successfully for several railroad section contracts, won them, and next established a reputation for compliance, good work and honesty. He also treated well the thousands of workers who, at different times, worked for him; he came to be respected by them, if not loved.

After a successful career in Chile, Meiggs went to Peru, in 1868. In Chile, he had built the line from Valparaiso to Santiago, and he was in the midst of finalizing the project for the Santiago–Buenos Aires line when General Pedro Diez Canseco became President of Peru and Dr Juan Manuel Polar his 'Chief of Cabinet' – an early version of Prime Minister. Polar had previously been the ambassador of Peru to Chile when Meiggs had met him and they had become friends. Polar was a railroad bug and now invited Meiggs to apply his talents to building railroads in Peru.

And that is exactly what Meiggs did. Within four months of his arrival, he obtained a contract to build the Mollendo–Arequipa Railway. Meiggs kept his promises and fulfilled his contractual obligations: the line was finished in three years. There was, admittedly, much talk of bribes, unscrupulous methods, and inflated charges, but what really mattered was that the line was built and that Arequipa was now in direct contact with the sea and with overseas markets.

Meiggs brought Chilean and Bolivian workmen to Peru. He put 12 000 men to work; 9000 of them were foreigners because there were not enough Peruvians with experience available. Yellow fever, smallpox and malaria took their toll; some 2000 died. But when the line was inaugurated, all the difficulties were forgotten. Some 800 guests came, including the new president, José Balta, and were treated royally. Meiggs knew how to impress people and make friends. Proudly he explained that the sleepers had been brought from Oregon, the rails from England, the locomotives from New Jersey, the coaches from New England – and much of the food the guests ate, from Chile.

A year earlier, Meiggs had obtained further concessions. One was to build a line from Lima to La Oroya. This was to be *the* railroad of western South America – connecting areas of almost unlimited mineral wealth and agricultural potential with the Lima market and overseas buyers.

Another concession was for the line from Arequipa to Puno, on Lake Titicaca.

Meiggs tackled both jobs with his proverbial gusto. He hired yet more Chileans and Bolivians and even brought some 6000 coolies from Macao and from the US, where they had helped to build the transcontinental railroad and now were just a jobless and unwanted – as well as very foreign – element.

It is difficult to find words to describe adequately the really gigantic and unique work that was done in those years to get the Lima–La Oroya line up into the mountains. Sixty-one tunnels and sixty-five bridges had to be built. But the most spectacular part of the job was the zigzag system of switchbacks used principally at the beginning of the line – near Chosica – to overcome the first and almost vertical ascent into the Sierra. A vast network of auxiliary roads and paths had to be built to get men and materials up the cliffs.

It should not surprise that the simultaneous building of two major railroads strained the resources of the Peruvian government to the limit. Actually, there was still more building going on. Meiggs also laid, during those years, 1870–3, the Chimbote–Huallanca (1972), Ilo–Moquegua (1873), Pisco–Ica (1873), Lima–Huacho (1870), and Pacasmayo–Guadalupe (1872) lines.

All five admittedly were small and technically not difficult, but they still required attention and swallowed up resources. Yet even that was not enough.

Though the Arequipa–Puno line was only finished in 1874, work on the Puno–Cuzco extension was already begun in 1872.

By 1875, Peru ran out of money and all work on the railroads was stopped. Meiggs and the government had bitten off more than they could chew. President Balta had been succeeded in 1872 by Manuel Pardo y Lavalle, who inherited the due dates and debts borrowed by Balta. Taxes covered only 50% of expenses and payments, and Peru was facing bankruptcy. Public works were stopped; the till was empty.

It was many years before the main railroads were finished. Although the roadbed of the Puno-Cuzco line had been completed, the rails were only laid years later and traffic inaugurated in 1907. The Central Railway was functioning to Chilca when work was suspended in 1875. The line reached La Oroya only in 1907.

With all work on his projects suspended, Meiggs struggled to find a way out. All he could think of was to put together a big mining company which would work the Cerro de Pasco copper and silver deposits – an idea he had probably had much earlier and which probably had been one of the driving forces behind the Central Railway project. Then the exploitation rights to the deposits could be sold for money with which to complete the railroad projects – plus a new one, from La Oroya to Cerro de Pasco.

The government actually agreed and Meiggs set out to print and sell his own paper money – 'Meiggs' bills'. But the public was sceptical. Even when he managed to get the government to guarantee them, distrust and rejection continued.

Then Meiggs suddenly collapsed. A stroke left him partially paralyzed. A few months later, in September of 1877, he died, at age 66. His 'empire' folded, work on the railroads remained suspended. A weak Peru struggled on to stay afloat – to then, two years later, in 1879, engage in a futile struggle with Chile, the so-called 'War of the Pacific', or 'Nitrate War'. Peru lost it – and the continuation of its railroad building programme got delayed still further.

And the opening up of the mines of central Peru was also delayed by at least ten years.[1]

Nowhere else in the Americas did railroads play as significant a role in the development of the mining industry as in Peru and Bolivia.

When the 1849 California gold rush came, tens of thousands travelled west – some across the continent (by rail as far as they could and then by coach, or horse, or on foot); others by sailing vessels around South America; still others by crossing Nicaragua; and still others by crossing the Panama isthmus – by mule, horse, or on foot. The idea of a railroad from the Atlantic side to the Pacific side was logical, and it was indeed built 1854–1856, specifically to serve the transport of people and goods for the California diggings.

The railroad soon paid dividends, 44% the first year. All of California's gold is said to have travelled this route.[2]

In Mexico, railroad building was late and slow. By 1876, a mere 400 miles existed. But the 'Age of Porfirio Diaz' was also good for railroads. By 1911, 11 800 miles had been built. No wonder mining took off. In 1990, the state acquired two thirds of the existing system.[3]

The first Canadian transcontinental railroad was completed in 1885.

And in the US? The 'push west', and the building of the first transcontinental railroad (completed in 1869) were principally to establish transport systems for grains and cattle, and for general geopolitical reasons, i.e. to establish links to the West Coast and to 'take possession' of the land. Once the lines were in place, both in the US and in Canada, they also became important for the growing mining industry.

Notes

1. The reader is referred to Bryan Fawcett, *Railways of the Andes*, a very informative and delightful book which reviews the history of railroading in the Andes in the general context of the times and of each country.
2. Arciniegas, *Biografia del Caribe*, pages 491–2.
3. Cheetham, *A Short History of Mexico*, page 201.

CHAPTER 11

Paraguay, Uruguay and Argentina

There is not a great deal to be said about mining in these three countries.

Paraguay

No major mineral finds have ever been reported in Paraguay, although a good part of the country is still covered by little explored savannas and jungle, and the discoveries of many tin, gold and wolframite placers, as well as bauxite and iron ore in Brazil in recent years have shown that surprises are still possible. Furthermore, not far away, near Corumba on the Brazilian side of the Paraná river, several deposits of iron and manganese ore have also been worked.

The only known, and then much touted, find of ore in Paraguay concerned tungsten in the 1950s. Philipp Brothers managed to buy the first lot, despite warnings that little or nothing was known of the mine. But those were the days of the Korean War, and much demand for strategic materials – of which tungsten was one – and securing supplies was most important. But, alas, the shipment turned out to be mere sand. It was a fraud, one of the few known to have been perpetrated on Philipp Brothers.[1]

Uruguay

Uruguay is also not exactly known for minerals. Still, in the 1980s, this small country became, for a while, a notable exporter of gold.[2] In this case, the metal did indeed exist; what was exported was indeed gold; only it was not Uruguayan gold, or not gold from Uruguayan mines. What was perpetrated was a different kind of fraud – smuggling. Brazilian gold was being smuggled across the border and then exported from Uruguay. The reason? To avoid income taxes, production taxes and export levies in Brazil.

By 1991, curiously enough, a gold mine had been discovered, near a town called Corales, and a Brazilian group identified as CMA-CMP, with some gold mining experience in the Amapa Territory of Brazil, near the border with French Guyana, will work the ore body. They claim to have 9 tonnes of reserves 'measured', plus another 10 tonnes 'inferred'.

But, vague as the information about the operation seems, the operators appear to have good partners, or owners, as Degussa of Germany, Monteiro de Carvalho of Brazil and TUC of Canada have been cited.[3] Maybe Uruguay will yet become a meaningful gold producer!

Argentina

Argentina, by contrast, does have something of a mining industry.

As early as 1516, Juan Dias de Solís reached the delta of the Paraná river and called it the Rio de la Plata, the silver river (in English archaically called the River Plate). A few years later, in 1524, Aleixo Garcia sailed further up the river and reached Paraguay and Bolivia. He did find silver and brought samples back.

Still another few years later, in 1526–8, Sebastian Cabot explored the delta further, to be followed by Pedro de Mendoza who, in 1536, founded Buenos Aires. But as he and his companions found no mines, the venture was a failure. In 1537, Mendoza returned to Spain; his successor, Juan de Ayolas, travelled up the Paraná river and founded Asuncion. Alvaro Nuñez Cabeza de Vaca discovered the Iguaçu waterfalls. Still they found no silver – but the estuary of the river retained its imposing name.

Some gold and silver were eventually found and the mines worked, for a while, during the colonial period. Some more production came forth between 1895 and about 1930.

The Aguillar deposit, in the Jujuy province, was discovered by Spanish Jesuits, but it really only began producing lead, zinc and silver in about 1936. Some gold comes from a mine called Farallon Negro, in Catamarca, which also has some silver production.

Argentina produced recently about 1 tonne of gold, some 70–80 tonnes of silver, 20 000–30 000 tonnes of lead concentrates and 30 000–40 000 tonnes of zinc concentrates – all per year. It has also produced about 1 million tonnes of iron ore per year.

Still, one wonders what made the people in some long gone and forgotten moment in time decide that the country, or the area or region should be called 'land of silver' – Argentina – and the mouth of the river, the silver river – Rio de la Plata.

Notes

1. A charming vignette, well remembered by those in the tungsten business and by former employees of Philipp Brothers.

   Some readers may wonder who Philipp Brothers are, or were. From very small beginnings in Hamburg, Germany, in 1901, as 'Julius Philipp', the firm grew and grew to become, after World War II, the largest and most powerful metals, ores and other raw materials (including oil) trader in the world. Between 1960 and 1980, this worldwide conglomerate pretty much dominated the metals scene and came to be called, not without reason, the 'supertrader'.

   Its decline began in about 1982, and was quite rapid; the company had ceased to function in 1990. By then it was a subsidiary of Salomon Inc; it remains today a dormant company.

   Why the firm declined and disappeared is still being discussed and may never be understood. Market conditions changed and this affected the firm's business – but such changes had occurred previously and repeatedly and the firm had always adapted to such changes and had thrived in adversity.

   In 1981, Phibro Corporation, the then parent of Philipp Brothers, bought Salomon Brothers, the money trader. As the result of several subsequent reversals, flipflops, and management changes, Salomon got the upper hand. It begs credulity that Salomon wilfully destroyed Philipp Brothers – as was privately alleged by former Philipp Brothers people. Salomon in time simply closed the non-profitable parts of Philipp Brothers and split off Phibro Energy, the profitable oil trading and refining firm.

   Would management incompetence explain the decline and demise of Philipp Brothers? Perhaps. But the quality of management is an intangible, difficult to gauge part of a business. Good management obviously makes a company grow and prosper, bad management . . . But then, luck also plays a role.

2. In 1987, Uruguay exported a staggering 60 tonnes. The information comes from an article by Diana Kinch, the Latin American editor of *Metal Bulletin*. The article was entitled 'Gold trading takes off in Brazil', and can be found in *Futures and Options World*, January 1989, page 21.

   Ms Kinch also told me that according to Nathan Blanche, President of ANORO, the Brazilian Asociacao Nacional de Ouro, Uruguayan exports of gold to the United States alone were as follows (figures in tonnes):

1985	36.0
1986	24.5
1987	10.3
1988	7.0
1989	0.7
1990	0.5

3. From an article by Altamir Tojal, entitled 'Passo Gigante – Group CMA-CMP salta do Amapá ao Uruguai para produzir 2.8 toneladas de Ouro', from the magazine *Isto É-Senhor*, March 1991.

CHAPTER 12

Brazil

Non-mining beginnings

Pedro Álvarez Cabral discovered Brazil for the Portuguese[1] in the year 1500. It appeared to be a beautiful and bountiful land – but not, apparently, with a rich and developed civilization. There also did not seem to be any mineral or metal wealth. And, indeed, as the years went by and the Spaniards opened up the wealth – first of the local civilizations and then of the mines – of Mexico, Peru, Colombia and Bolivia, the land 'of the True Cross' (Vera Cruz) quickly to become known as 'of the Holy Cross' (Santa Cruz), and then to be renamed Brazil, did not live up to much more than some meagre returns from brazil wood, a colouring agent from which the country got its name. Yes, there were also monkeys and parakeets (*papagaios*) but both were of absolutely no value to the gold and wealth-hungry Europeans.

Subsequently, after the brazil wood trade declined, it was from about 1530 sugar, mostly from Pernambuco, that constituted the economy of the colony. Metals were discovered only much later – but then in substantial quantities and over much of the area between Bahia and the River Plate. Brazil was thus almost a full 200 years behind Mexico and Peru (at least when seen with European eyes) and, in fact, several thousand years behind the pre-Inca discoveries and workings of gold and other metals.

Sugar, originally of Far Eastern origin and a luxury item in Europe purchased at great expense from the Italian (Venetian, Genovese and Floretine) traders, eventually came to be produced in Sicily. Later, by the year 1498, Portuguese sugar from Madeira dominated the market. But soon it was production from Hispaniola (after 1505) and then from Pernambuco (after

1526)[2] that shifted the supply sources west. By 1560, Brazilian sugar was an important factor in the business,[3] and, between 1600 and 1650, the Portuguese even dominated the world sugar market with their large Brazilian production.[4] From about 1670, however, competition from Dutch, British and French producers in the Caribbean caused the Portuguese difficulties in disposing of all of their Brazilian sugar.[5]

But, contrary to the other American colonies, there were no metals in Brazil – or so it seemed.

The pre-Columbian Brazilians were Stone Age primitives. No civilization worth that name was found by the Portuguese. And no metal artefacts – gold or silver jewellery or copper tools – were found, at least not in any quantities nor of quality comparable to the artefacts of the Inca, Aztec, and Chibcha civilizations and their predecessors.[6]

But the metals – and minerals – were there as later discoveries would show. No silver, but gold in large quantities, some copper and, to be discovered and extracted much, much later, diamonds, iron ore, tungsten, bauxite, manganese ore and large quantities of tin.

Gold in Minas Gerais . . .

Gold was the first of the metals to be found – alluvial gold in the rivers and in placers nearby. The time of the first major gold rush (the second took place in the 1980s) was the late seventeenth century, although the largest production came only between 1741 and 1761.[7]

In about 1560, one Brás Cubas (he who founded Santos) found some gold near the modern city of São Paulo,[8] as did Afonso Sardinha at what must have been a sizeable mine in Jaraguá (now a suburb of São Paulo). The quantities were large enough, or at least the production was steady enough, to cause the setting up of several small foundries, in São Paulo (about 1614), Iguape, Paranaguá and, somewhat later, in Taubaté.[9] They must have been small melting shops, for melting down and casting into bars the fine gold dust and the few nuggets. These melting shops were important for the government – to collect the Royal Fifth.

Still, it was not much. A few nuggets found here and there, some dust washed out of the riverbeds – it did not make for a gold rush.

Up in the north-east, in Bahia, Pernambuco, Sergipe, Paraiba, Ceará and Maranhão, practically all economic activity revolved around sugar. Much of the technology apparently came from the Dutch who between 1624 and 1654 ran a colony, through the Dutch West Indies Company, in Pernambuco. Growing sugar cane and producing sugar were profitable – as long as sufficient numbers of black slaves arrived from Africa, to replace those that died and to allow expansion. And profitability with sugar caused no push – out of boredom, need or curiosity – into the interior.

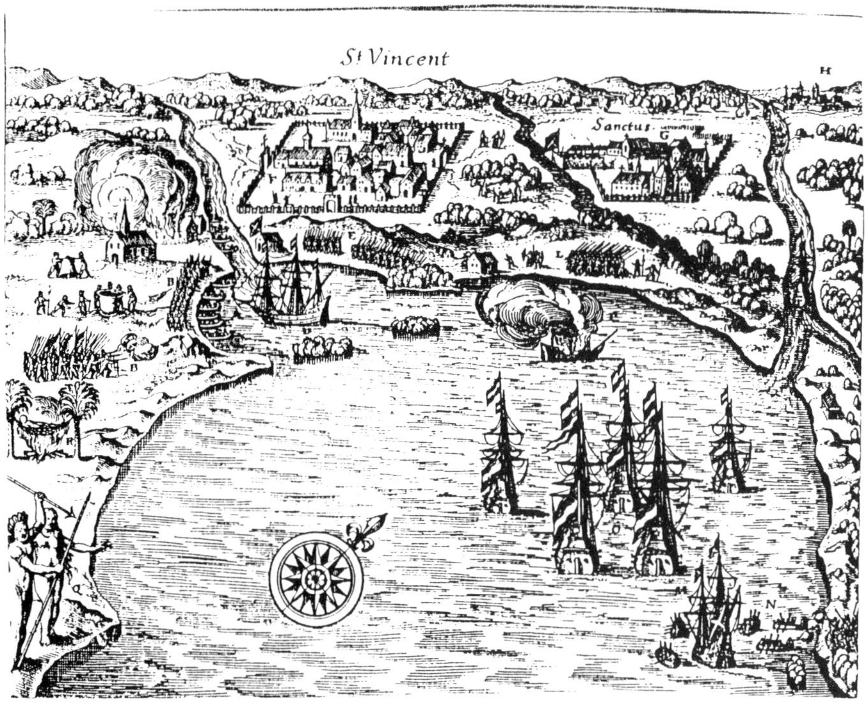

A view of the port and town of São Vincente, in about 1614. In time the port became Santos. Source: Johan Ludwig Gottfriedt, *Newe Welt und Americanische Historien*, first published in 1631 by Matthaeus Merian dA and republished in 1980 as Stuttgarter Faksimile Edition im Fackelverlag, Stuttgart, Germany.

When the push into the interior eventually did come, late in the seventeenth century, it was caused by several factors.

To begin with, sugar values were by then declining. By 1654, the Dutch had been expelled and had set up rival industries in the Caribbean from where cheap sugar – of Dutch, French and British origin – competed successfully with Portuguese sugar from Brazil. Other products had to be found by the Portuguese in their colonies.

Also, population pressure must have been felt in São Paulo, less from too many people than from a lack of jobs, as no major industry has as yet been developed in the São Paulo area.

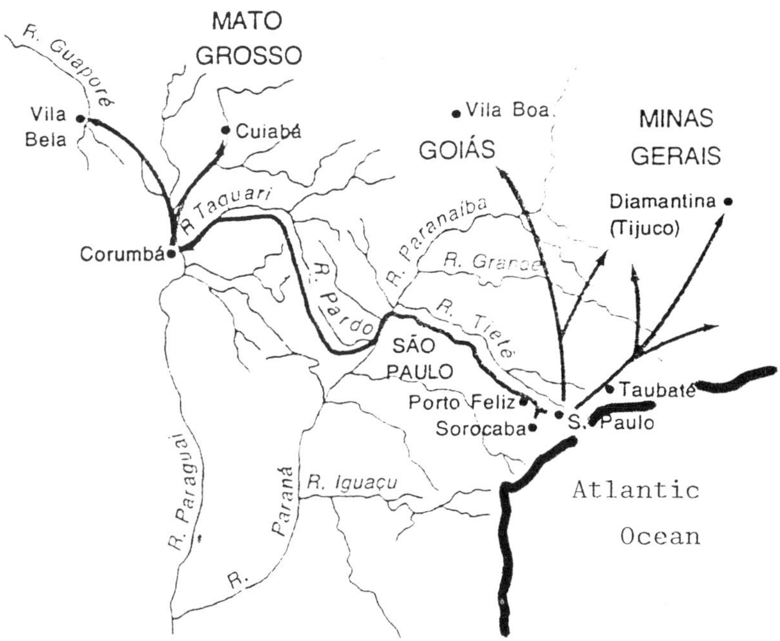

The Brazilian gold rush of the eighteenth century: the push from São Paulo into the interior, to Minas Gerais, Goiás and Mato Grosso.

Thus, when the Portuguese crown called for renewed efforts to search the interior for mines, *bandeiras* – bands of men, adventurers, following a leader and a flag (a *bandeira*) – set out, on foot, by canoe, even sometimes on horseback, to explore, conquer and eventually settle the interior. Most of the early *bandeiras* set out from São Paulo; they moved north and north-east, into Minas Gerais, and also, later, into what would in time be called Mato Grosso and Goiás.

Somewhat earlier, in 1637, one of the *bandeiras*, under the command of Pedro Texeira, had set out with 2000 men from Belém, to go up the Amazon, Silimões and Napo rivers, eventually reaching Quito in Equador. It had been the first ascent of these rivers.

In 1648, Antônio Raposo Tavares marched westward from São Paulo into Paraguay, to then turn north, eventually descending the Madeira and Amazon rivers to Belém. Later, other *bandeiras* followed, mostly in pursuit of slaves and in search of gold. They founded the cities of Paranaguá in 1648, Florianópolis in 1678, and Curitiba in 1693.

The most famous of the *bandeirantes* and really the first to set out in pursuit of gold and gemstones was Ferñao Dias Pais who between 1674 and 1681 repeatedly crisscrossed Minas Gerais, accompanied by other well-known figures such as Garcia Rodrigues and Manuél Borba Gato. In 1694, one Arzão found gold in Caeté, and his brother-in-law Bartolomeu Bueno de Siqueira soon exploited these deposits. Other finds soon followed, in 1698–9, on the banks of the das Velhas river, one of the tributaries of the São Francisco river.[10]

The adventurer Borba Gato, whose troops a few years earlier had murdered Rodrigo de Castelo Branco, a royal official, now made his peace with the authorities and emerged from the interior. He revealed to the crown – in exchange for immunity – the rich deposits of Sabarabuçú.[11]

The population of São Paulo – and the area around it and all the way down to the coast – cannot have been large. And it seems that the population of Minas Gerais was practically zero in about 1695. By 1709, it was up to 30 000, and by about 1795 to about 500 000 people.

And from nothing, in 1695, sprang the principal mining towns of Sabará, São João del Rei, Ouro Preto and Mariana.

In less than 20 years, between 1695 and 1712, a major gold rush took hold of Brazil – first from São Paulo, going north into Minas Gerais, but soon also from Rio de Janeiro, and finally going west from Vitória in Espírito Santo and south from Bahia. Many Portuguese came over from Portugal to join in. And soon they were at each others' throats, the locals, or Paulistas, and the new immigrants. The latter eventually won and the Paulistas looked for gold elsewhere.

By 1720, Minas Gerais became a separate captaincy.

The history of the Minas Gerais gold rush is colourful – but in no way different from other gold rushes. Many tried their luck; some placers were very rich, most mediocre or poor. Those prospectors that 'made it' soon built houses – and churches – in the cities of Ouro Preto, Mariana, São João del Rei and others, the splendour of which can still be admired today. Modern Brazil has done a respectable job of reconstructing and preserving these baroque towns.

The Brazilian gold rush, or 'Age of Gold' lasted more or less for a century – in fact, not even for that long. The important years are generally considered those between 1700 and 1770 but one can, of course, also include more of the colonial period, the years between 1650 and 1820.

Various historians have estimated what the gold production may have amounted to. The numbers are confusing, and striking a balance between them or taking an average of the many numbers does not really answer the questions.

The fact is that the really meaningful gold rush took place between 1720 and 1800 when about 10 tonnes per year were produced. Before 1720, production must have been less than 2 tonnes per year for much of the seventeenth century; quantities before 1630 appear to have been negligible. After the year 1800, production again fell to about 2 tonnes per year. This would give a total of 1180 tonnes, i.e. 180 tonnes between 1631 and 1720, 800 tonnes between 1721 and 1800 and 200 tonnes between 1801 and 1900.

But much of this is guesswork. The most important question is how much production was smuggled. All numbers that have come down to us are 'net' of contraband or include, at best, a factor of 20%.[12]

For most of the colonial period, the Royal Fifth applied, and many of the statistics are extrapolations from the tax registers. But one-fifth, or 20%, was a high tax – and the higher a tax, the greater the incentive to cheat. We can thus safely assume that much contraband took place and that the real production of Brazil in the colonial period may well have been much higher – maybe as high as 2000 tonnes between 1600 and 1820. On the other hand, the Brazilian ports were rather closely controlled by the Portuguese crown and not much contact existed with traders from other nations, neither via the ports nor through overland traffic to Peru, Equador, Bolivia and Argentina. The distances were enormous, the geographic obstacles formidable and at the other end loomed more tax collectors. Thus, speculation about what the gold production in the colonial period was really like will continue.

One of Brazil's foremost historians of the economic patterns and problems of the colonial period, Roberto C Simonsen, has speculated that Brazil produced, between 1700 and 1770 – the prime gold production time – as much gold as the rest of America put together produced between 1492 and 1850.

Almost all of the gold of the early years came from placers. Little deep rock mining took place. It was only in the middle of the nineteenth century that a few real mines began to be operated such as, from 1830, the São José and Saõ João mines by the St John d'El Rey Mining Company Ltd; and from 1834 the company also operated the Morro Velho mine at Nova Lima, near modern-day Belo Horizonte. The mine had by 1990 produced about 250 tonnes of gold, and is still operating – see below.

Baron Wilhelm Ludwig von Eschwege, who was in Brazil in the early part of the last century and who is often called the father of Brazilian geology, calculated, back in 1830, that the quantity of gold produced up until then (about 950 tonnes, including some 150 tonnes of contraband) had originated as follows: 68% from Minas Gerais; 17% from Goiás; 6% from Mato Grosso; and 9% from São Paulo.

But the old – historical – quantities of gold, meaning the estimated production of 2000 tonnes between 1600 and 1820, including maybe as much as 15 tonnes a year in the best period, 1735–51, fade into insignificance when compared to modern production figures – about 100 tonnes per year in the early 1990s.

During the early years of the gold rush, until about 1715, most of the gold

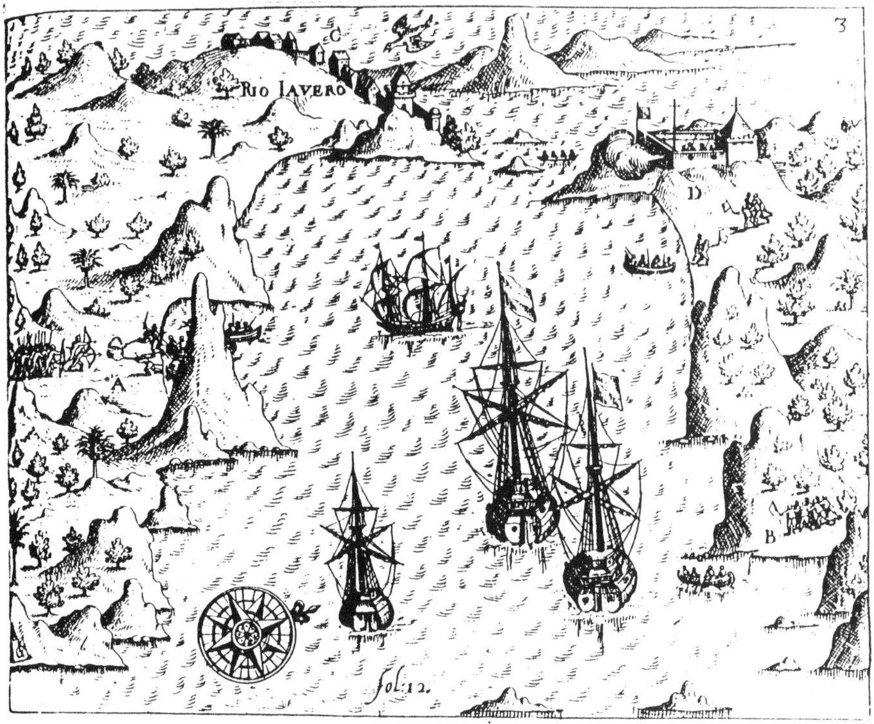

A view of Rio de Janeiro in around 1598. Source: Johan Ludwig Gottfriedt, *Newe Welt und Americanische Historien*, first published in 1631 by Matthaeus Merian dA and republished in 1980 as Stuttgarter Faksimile Edition im Fackelverlag, Stuttgart, Germany.

came out of Brazil via a route that would today lead us from Ouro Preto through the towns of São João del Rei, Baependí and Quelúz down to Paratí or Angra dos Reis, the old, small ports. Or via Taubaté where the melting shop and tax office were.

But the route was considered much too long and too much under the control of the Paulistas. Thus, from about 1715, alternative routes were tried, essentially from Rio de Janeiro into or through the Guanabara bay, by boat to the port of Pilar, and then via Tungua to Paraiba do Sul, and then more or less straight north to Ouro Preto – or, vice versa, from Ouro Preto down to Rio de Janeiro with gold.

From about 1720, when the new route was first improvised and then used extensively, Paratí began to fall into disuse. It was only once gold began to flow out of Goiás, after 1726, and then out of Mato Grosso that the port came to regain some of its former shine. But the bulk of the traffic, gold moving down from the serra (the Portuguese word for sierra, or mountains, but here really meaning 'the hills of Minas Gerais') and merchandise up, flowed via Rio de Janeiro, and it made Rio a rich and important city, the capital in fact as from 1763, when the administration was moved to Rio from Salvador da Bahia.

. . . and elsewhere

Goiás

It must have been in about 1682 that Bartolomeu Bueno da Silva, nicknamed Anhangüera, also O Velho, marched into Goiás – probably from São Paulo through western Minas Gerais. It must have been a long and arduous trek. The objective was Indians. It was a slaving expedition.

We do not know how many slaves they brought back, but Bueno da Silva found gold, somewhere in central Goiás, beyond the Paranaiba river and today's city of Goiánia (which was only planned and built between 1933 and 1942). Already prior to Bueno da Silva's expedition, one Manuel Correia had brought a handful of gold, some 35 grams, from the area.[13] Now, in 1682, Bueno da Silva brought back 150 grams – and a description of natives, of the Goiás tribe, covered with leafgold.[14]

In 1726, Bartolomeu da Silva's son set out once more, and he struck it rich – in relative terms. When he returned, he brought at least 30 kilograms.[15,16] The gold had apparently been washed at a place later called Porto do Meio, where the Vermelho river joins the Bugres river, near today's Goiania. Some of the finds were quite rich as a man could wash, or produce, about 12–15 grams per day. One pan, in fact, gave as much as 250 grams.[17]

Some of the placers were apparently revealed by the Indians – women taken prisoner who then took the white men to the mines.[18]

Food was scarce, at the placers. All that everyone wanted was to wash gravel and get rich quick. But some of the merchants were smart. They brought goods – presumably by mule train – from São Paulo and cashed in handsomely. A measure of corn went for 25 grams of gold, a pound of sugar for 7 grams.[19] When a few of the local residents tried their luck at planting something, they also reaped good profits.[20]

The law was far away; in fact, there was no law except that of the jungle, or of the strongest. One woman was said to have murdered her two sisters out of jealousy over men; she was also said to have suspected her slave girl of being the fruit of her own lover's loins, whereupon she killed the slave, impaled her and served her, roasted, to her lover.[21]

Washing river gravel for gold in north-eastern Brazil (by Percy Lau). Source: reproduced by permission of Revista Brasileira de Geografia and of the Fundacao Instituto Brasileiro de Geografia e Estatistica – IBGE.

Bueno da Silva the Younger, the principal discoverer of the Goiás placers, and by now a wealthy man, appealed to the governor for 'more', presumably titles and tithes. When he was rebuffed, he resolved to send his son, Francisco, and his son-in-law, João Leite da Silva Ortiz, to Lisbon to apply to the king.

The two took 30 kilos of gold, for the expenses of the trip, and set out down the São Francisco river to Bahia, thence to Pernambuco, and on to Lisbon.

But Ortiz died on the ship, and Francisco Bueno also died soon afterwards in Lisbon. And the 30 kilograms of gold? They were lost; no trace was reported.[22]

More placers were found, again and again, throughout the centuries, and well into our times. By 1989, the northern part of Goiás was separated from the state to form a new state, Tocantins. The rivers, and dry riverbeds, are still being washed.

Mato Grosso

The story of gold in 'the great jungle' (Mato Grosso) was no different. Pascoál Moreira Cabrál is credited with finding the first placers near – actually right in – today's Cuiabá, in 1719, or maybe a year or two earlier; 1719 is the year when the find was registered. In those days it took seven months by canoe to get there.[23]

The Cuiabá deposits were shallow and not very rich.[24] By 1732, they were supposedly already exhausted, having yielded 'several hundred arrobas' (3.5–5.0 tonnes) of gold.

Also in Mato Grosso, placers have been worked off and on throughout the centuries, and both states, Mato Grosso and Mato Grosso do Sul, continue to be producers of substantial quantities of gold, even now in the 1990s.

Bahia

Gold was also found in Bahia, in 1718, by Sebastião Pinheiro Raposo, along the Rio das Contas.

More about gold in the Amazon basin in the 1980s and 1990s follows on pages 192–7 and 199–205.

Diamonds

Xica da Silva, a mulatta in her early twenties, must have been an interesting woman. She persuaded her lover, João Fernandez de Oliveira the Younger – who between 1759 and 1771 held the government monopoly exploiting the diamond mines of Tijuco, in Minas Gerais – to build her a lake, large enough to hold a sailing ship. She simply wanted to see a sailing ship once!

And the wealth to build such a lake, the ship, plus country homes and other fineries – all came from diamonds.

Tijuco still exists, but it is now called Diamantina. It is located some 260 kilometres north of Ouro Preto.

The gold prospectors who had poured into Minas Gerais late in the seventeenth century, and early in the eighteenth, had not stopped in the valley of Jequetinhonha and around Ouro Preto, Mariana, São João del Rei and Sabará. They moved on. Of course, those who had found rich deposits, had staked out claims and were producing, stayed. But the many who came too late, or were unlucky when panning the rivers, moved on, mostly north.

They followed the streams and rivers, and panned – and eventually, in around 1726, somebody found diamonds. They did not recognize them as such. It took experts in Lisbon to identify the glassy stones. And from about 1728, a new rush was on.

For nearly a century and a half, Brazil then became the leading world producer of diamonds. It replaced India, which until then had been the world's leader, but whose deposits were becoming exhausted.

Brazil remained the world's principal producer until about 1860 when the much richer mines in South Africa were discovered.

Still, although Brazil is not any more the leading producer of diamonds, it is today probably the country with the greatest variety of precious and semi-precious stones.

Back then, in about 1728, one Bernardo da Silva Lobo took a few stones to Lisbon where they were seen, quite by accident, by the Dutch consul who identified them as diamonds. Da Silva, who was henceforth acclaimed the discoverer of the Brazilian diamonds, was richly compensated by the king.[25] But for quite some time thereafter, the Brazilian diamonds had to be shipped to, and sold through, the Portuguese dealers in Goa – then a Portuguese colony – and had to be traded on the established reputation of the Indian stones.

Much earlier in Brazilian colonial history, Ferñao Dias Pais, the one who between 1674 and 1681 crisscrossed Minas Gerais, had found 'diamonds'. They were at first called 'coloured stones' (*pedras coradas*) and, then, emeralds. But they were most probably green tourmalines. And throughout the centuries more and more different types of semi-precious stones were found – mostly in Minas Gerais and Mato Grosso, but also in other regions of Brazil, as far north as Ceará (tourmalines and aquamarines) and as far south as Rio Grande do Sul (agates, amethysts, citrines and opals).

The Brazilian diamonds were (and most still are) found in river gravel. No 'pipes' (the vertical, cylindrical matrix, of eruptive origin, in which diamonds are found in South Africa) have ever been located in Brazil. These are said to be the 'matrices' from which or through which diamonds came from the inner earth.

The extraction methods were basically similar to those of alluvial gold – washing gravel to get the glassy diamonds.

How much was ever produced? Throughout the colonial boom period, about 1728–1810, some 3 million carats were exported – so the statistics say.[26] But how much was smuggled? It is anybody's guess. The authorities were extremely tough and determined to prevent fraud or smuggling. After all, the crown and the authorities needed their taxes. But the operators were also tough. Many of the slaves – because the diamond industry, as with panning for gold, depended heavily on slave labour – were often force-fed a purgative, to 'wash out' any diamonds swallowed during work.[27]

Work was at first done simply by *garimpeiros* – those wild prospectors – extracting the diamonds freely, alone or with slaves. Then, from 1740 to 1771, contractors got exclusivity, for payment of fixed yearly fees to the crown. But from 1772 to 1800, the crown exploited the deposits directly.

More than 40 000 miners rushed to Tijuco and the surrounding areas, all in search of diamonds. Effective poiicing was obviously impossible – and the major part of what was found was no doubt smuggled out. So much was, indeed, produced that the price of diamonds in Europe fell,[28] and substantially.

But all the old numbers are incomplete. Forty thousand men – of whom most must have been slaves – would have given the crown an income, through a fee, or poll tax, raised rapidly to 20 000 milreis per worker, a very substantial amount. But no such income was registered, neither in Minas Gerais, nor in Lisbon.[29]

When it was perceived that the system did not work, the order was given to kick all prospectors out of the area. But a few months later, the governor relented and let them stay.[30] Then, in 1732, the governor decided all blacks and all mulattos should be expulsed, but the order was apparently ignored.[31]

In 1733, the next governor increased the tax from 20 000 to 25 000 milreis per worker. He also forbade all 'beggars and vagabonds' to enter the district, and he closed the taverns at night.[32]

Not much, it appears, has changed in 250 years; today, modern Brazil also prohibits, by law, the participation of foreign individuals in gold mining in the Amazon – and also the sale and consumption of alcoholic beverages within 1 kilometre of all major highways. Both laws, like many others, are merrily ignored, as were, apparently, those issued in Minas Gerais in about 1733.

Late in 1733, the poll tax was raised to 40 000 milreis. This means that between June of 1730 and December of 1733, it had risen from 5000 to 40 000 milreis – an eightfold increase.

Diamond mining was either extremely profitable, to absorb such tax increases, or nobody paid and the rate was thus irrelevant. And the colonial government, as with today's Brazilian administrations, raised its expenditures heftily, and then raised taxes with equal abandon. But, and once more as today, exorbitant taxes drove business, and income, into illegality.

Only a few months later, yet another decree abolished all taxes, replacing them with the obligation that all diamonds larger than 20 carats belonged (and were to be delivered) to the crown. And there was now a tax of 60 000 milreis on each store, and of 36 000 milreis on each tavern.

Yet another governor reinstated the poll tax and raised it, in 1739, to 230 000 milreis. But this time the governor had outreached himself. Work stopped, as nobody was willing, or able, to pay such an exorbitant fee for each labourer.

An all-embracing concession, and monopoly, was now given in 1740 to João Fernandez de Oliveira. It was initially granted for four years, a period which was then extended for another four years. In 1748, one Feliberto Caldeira Brant got the concession until 1753, but Fernandez de Oliveira got it back in 1754. With two partners he held it until 1771.[33]

But the crown apparently got the short end of the deal and decided in 1771 to henceforth once more operate the mines directly.

How much was smuggled out we do not know. As stated already, several calculations put the quantity of diamonds from Brazil, between 1740 and 1828, at something like 3 million carats. Would the real quantity have been double, or triple, or 10 times that number? The reader is left to muse. Between 1832 and 1936, more than five times as many stones were mined in Brazil.[34]

By 1980, Brazil's production was estimated at 350 000–400 000 carats a year.[35] By 1990, it was probably at 500 000 carats.[36] These figures are estimates, not official numbers, and do not differentiate between what is sold legally and what is smuggled.

Emeralds were finally discovered in 1963, in Salininha, in the state of Bahia. Aquamarines made the town of Teófilo Otoni, in Minas Gerais, famous, and there are many deposits of tourmaline and topaz and the many other semi-precious stones.

These semi-precious stones really only became important after World War II. Until then they had been largely ignored. It was during the war that American demand for quartz and mica – both then known to exist in Brazil – produced semi-precious stones often as by-products of the mining of these minerals then considered strategic.[37]

The Chapada Diamantina

There was another diamond rush from about 1844, in the very heart of the state of Bahia. In that year, superficial alluvial deposits were discovered in and near the beds of the Mucujé river, as well as the Lençois and San José rivers.

So rich were the deposits that within a few years a number of towns rose, among them, and principally, Lençois (which means 'sheets', or 'bed sheets', supposedly from the view the traveller got from the surrounding hills of innumerable 'sheets' spread over the workings). Soon Lençois progressed from a primitive mining camp to a town, with small, quaint and (still today) colourful houses, but also ruthless mining bosses and traders, many slaves and many whores.

So important did Lençois and the Chapada diamond region become, in the latter part of the nineteenth century, that the politicians in Salvador even considered moving the capital of the state to another of the mining towns, Cascavel, at the southern end of the mountain range. But the promise of the deposits petered out, the capital was not moved, and a railroad, often promised, was not built. And what in about 1850 or so, was supposedly a town with some 30 000 inhabitants (to which one might want to add another 30 000 or so *garimpeiros* or miners working the riverbeds and hills), became, by about 1930, an almost abandoned hamlet of only some 2500 inhabitants.

Production data do not exist, or cannot any more be found. Some speculation points to annual quantities of something like 60 000 carats, but we do not know for how many years this lasted, and of what quality the diamonds were.

The town of Lençois still exists, and offers proof, with its layout and beautiful buildings that quite some wealth must have existed. But today the city ekes out a meagre living with two or three small, mechanized (monitors and primitive dredges) diamond mining operations, and tourism. The mountain range itself has been declared a national park.[38]

Baron von Eschwege

One of the outstanding personalities in the history of mining in Brazil was Baron Wilhelm Ludwig von Eschwege.

Born in 1777 in Eschwege, Germany, he studied engineering and, once he had graduated, looked for employment beyond the confines of his home district, Hessen, and Germany. Thus he became, in 1802, at the young age of 25, 'Director of Mines' in Lisbon.

When Napoleon I invaded Portugal, Eschwege fought on the side of the Portuguese and, probably in recognition of his services for Portugal – but also for his technical qualifications – the Portuguese–Brazilian Prince Regent, Dom João, invited him in 1810 to travel to Brazil, where the Portuguese crown and court had settled since 1808. Eschwege was to become a high official in the Royal Corps of Engineers (of which he indeed eventually became a colonel) and director of the Royal Mineralogical Council.[39]

He was soon sent to Minas Gerais, in 1811, and settled in Vila Rica de Ouro Preto. His task was to make a mineralogical survey, to improve mining methods, to increase gold production, and to investigate the possibilities of shipping on the Doce river. The source of this river is close to Ouro Preto from where it makes a wide sweep north and then east before it reaches the Atlantic Ocean. The valley the river forms, the Vale do Rio Doce, became known later as one of the richest iron ore bearing strata in the world; it was also to give its name to the Companhia Vale do Rio Doce, Brazil's big government owned iron ore producer.

And thus Eschwege, with German and Hessian stubbornness, thoroughness and persistence, travelled all over Minas Gerais, studying, advising and taking notes. What was there was of interest to him, as well as to history; he wrote it down, and in time published it in German in 1833 in Berlin. It was called *Pluto Brasiliensis* – 'the Brazilian Richness'.

Eschwege stayed in Brazil for 10 years, from 1811 to 1821. In 1812 he produced the first Brazilian pig iron; experimentally, yes, but still . . .[40] His survey of mining in Minas Gerais, his review of the gold occurrences, extraction, and the reasons for the decline of the industry, as well as his review of diamond and iron ore mining are still today basic for an understanding of the early history of mining in Brazil.

The Morro Velho gold mine

This mine is located a scarce 10 miles from the centre of Belo Horizonte, Brazil's fourth largest city. The township is now called Nova Lima, but it is actually Vila Nova de Lima; it used to be called Nossa Senhora do Pilar de Congonhas de Sabará.

The mine is called Morro Velho (Old Mountain), but others of the same complex are called Raposos, Bicalho, Faria, Bela Fama and Esperança. There are also Mina Velha and Mina Grande. It consists, in other words, of a series of mines.

Meaningful mining of the outcroppings began in 1830; it is now considered the oldest mine in Brazil, and the oldest worked gold mine in the world. It is claimed it is also the deepest gold mine in the world.

Little is known about the early history other than that the first workings were begun in around 1725. The company itself – now called Mineração Morro Velho SA – claimed in 1954 that the mine had been worked since 1750. A family by the name of Freitas then owned the operation.

When, in 1829, the further importation of slaves was forbidden and difficulties arose of how to get more, or new, labour, the Freitas family sold out to British interests, represented by a certain G Francis Lyon. This is how and when the St John d'El Rey Mining Company Ltd came about, owning at first the São José and São João d'El Rey mines. Regular production started in 1834.

The first workings were at road level and were simply holes hacked into the hill. In 1835, production reached only 99 kilograms; but by 1853 it was up to 1337 kilograms. By 1849, work had proceeded along the vein to a depth of 110 metres. The vein was then 1.4 metres wide. Some 1100 people brought out 6000 tonnes of ore per month.

A few years later, in about 1867, the ore was running 23.5 grams per tonne, and they recovered 15.5 grams per tonne.

We are allowed to satisfy our curiosity and to make a few calculations. If of the 1100 labourers 1000 worked with the ore, breaking it and transporting it, each would have handled 6 tonnes per month or at, say, 25 working days a month, 240 kilograms a day. If one-third of the men broke ore, one-third transported and one-third milled and processed it, each would have handled about 720 kilograms a day – not unreasonable.

And at 15.5 grams per tonne recovery, it would have given them 93 kilograms a month, or about 1.1 tonnes for a year – consistent with the old figures.[41]

By 1867, they were down to 335 metres. Then a great fire damaged major parts of the mine. No wonder: not only had huge quantities of timber been introduced – many logs were as long as 14 metres – but also much oil was used, for lamps and to combat termites.[42] In 1856, for example, some 200 000 candles were used.

They decided to sink two new shafts, and reached the ore body again five years later, at a depth of 318 metres: no mean achievement.

Production resumed in 1872 – although it had never really stopped as they had reworked earlier rejects and lower grade ores.

In 1886, most of the old morro caved in. Whereas the fire of 1867 had not taken any lives, this time several miners died.

It would have been easy to now give up. But they once more persisted and decided to sink two new shafts, 460 metres to the north of the centre of the mine. And 690 metres down they again hit the ore body – although it was only 4.5 metres wide! By 1892, production began once more, after an interruption of six years.

Over the years, they went deeper and deeper. By 1915 they reached 1777 metres, and by 1934 2453 metres.

As they went down, heat became a problem, rising two degrees with each 54 metres. Eventually ventilators with ever larger capacity had to be installed. As the temperature was bound to rise to 52 degrees celsius at 2300 metres depth, the installation of meaningful air-conditioning became mandatory – and was achieved by 1912. A second, better installation being manufactured in Britain was delayed during World War II. For a while, work had to proceed at 48 degrees celsius.

The greatest depth they ever reached was 2453 metres in 1934. They did not go deeper and, since 1969, work has continued at around the 2200–2300 metre levels.

Today Mineração Morro Velho SA produces around 6600 kilograms of gold per year (1990 figure), from ore that still runs 8 grams per tonne (run-of-mine). A new cooling plant installed in 1980 brought air temperatures down further, from 34 degrees to 27 degrees.

And while the owners – Anglo American and the Bozano Simonsen group – are understandably cagey about the reserves, enough gold-containing ore, of payable grade, seems to have been discovered to assure continuing operations well into the next century – if the gold price rises from recent levels of US$350 per oz. If prices do not rise, Morro Velho may well close soon.

A quiet 100 years, 1850–1950

Throughout its colonial history, Brazil was always a producer and supplier of raw materials to Portugal. And nothing changed when Brazil became a kingdom (in 1815), an empire (in 1821) and during much of the republic (since 1889). Brazil continued as a supplier of tropical goods. And as such, the country went through different cycles – brazil wood, sugar, gold, diamonds, cotton and rubber, as well as coffee – without ever setting up much industry.

After the rush for gold and diamonds subsided in about 1810, the country settled into this cyclical pattern.

Portuguese crown policy was not much different from that of England about its American colonies and Spain about its overseas possessions, i.e. to forbid all industry so that all manufactured goods would come from the mother country.[43]

This explains why no meaningful metal industry was developed in Brazil – and hence no meaningful mining industry came about – in the seventeenth, eighteenth, and even in the nineteenth centuries. We shall talk about the off-and-on iron industry in São Paulo later in this chapter.

When, fleeing from Napoleon's army, Dom João VI came to Brazil, he was accompanied by 15 000 people, the best of Portugal. In the years 1807–10 he suddenly opened up the country, allowing foreign vessels to enter the ports and trade. He decreed that manufacturing plants could be set up and he provided the incentive for the opening of the first Banco do Brasil. Still though, not much happened.

For a while still, gold figured as a meaningful export item – but total exports were in fact very low in those years (see Table 12.1). These figures come from Roberto C Simonsen's book, *Historia Economica do Brasil (1500–1820)*. But they are really abbreviations. The real figures, expressed the old way and copied straight out of Simonsen's book, look as follows, taking the first figure as an example: 5851:244$999. These figures are said to have been *réis*, the monetary unit in use then.[44]

But then in about 1820 gold disappears from the statistics. It had first been

Table 12.1. Exports to Portugal of Brazilian goods, 1796–1819

	1796	1809	1819
Food	5 851	6 535	4 591
Gold*	1 856 (16%)	940 (5%)	26 (3%)
Cotton	2 201	3 544	1 784
Hides	738	2 249	625
Drugs	163	285	60
Lumber	26	36	13
Tobacco and others	640	566	362
Total	11 475	14 255	7 461

* Percentage figures relate to gold value as a percentage of total exports.

Table 12.2. Brazilian exports, 1821–1900

Period	Product	Percentage of total exports
1821–30	Sugar	35%
	Cotton	24%
	Coffee	21%
1851–60	Coffee	54%
	Sugar	23%
	Hides	8%
1891–1900	Coffee	67%
	Rubber	16%
	Sugar	6% (expressed in values)

shipped out in 1690, had accounted for half of all exports in 1760, and now exports ended.

This does not mean that production of gold (and diamonds) ended. Some mining activity continued, but the production must have been small and was mostly absorbed locally – no doubt going into consumption underground.

We know with a great deal of certainty that the Brazilian exports of the nineteenth century included no metals and no metallic products.[45] The exports consisted of primary, and mostly agricultural products, see Table 12.2.[46]

By 1846, machinery could be imported duty-free and in 1851 the second Banco do Brasil was founded.[47] But these and other liberal laws and openings did still not cause much industry to be set up. The pull towards agricultural production – with the easy and large profits in coffee, sugar and cotton – prevailed. Slavery, only abolished in 1889, was replaced by cheap immigrant labour. Not even the introduction of steamships that cut the trip between

Belém and Manaus from two to three months to nine days, nor the building of railroads (from 1854 on) did much for industry.

As shall be seen below, it was principally the lack of enthusiasm to get into iron- and steel-making that kept the country back.

But then Brazil was, in those years between 1850 and 1900, still sparsely populated (5–15 million inhabitants) and, being a huge country, no industry seemed to make sense anywhere, for lack of consumers and almost total absence of means of transportation. To bring manufactured goods from Europe by ship, to Salvador da Bahia, or Belém, Recife, Rio de Janeiro as well as São Vincente (Santos), was much less complicated. And no major foreign threat existed to warrant arming and maintaining a large army (which would have involved setting up a large armament industry), and when a threat was perceived (for it in reality never existed), like in 1865 from Paraguay, it still did not bring about an armament industry. Borrowing money abroad and importing weapons was faster and easier.

The 'New Brazil' that emerged after the monarchy was overthrown, in 1889, did bring about fundamental political changes; yet the agricultural economy, basically feudal and characterized by latifundia and monoculture, continued. And the same pattern persisted until about 1960.

Changes did, of course, take place. The country industrialized slowly, especially as a result of World War I when, for a while, the flow of manufactured goods from Europe was interrupted, or at least curtailed. But Brazil's agricultural production continued to find outlets, and got better prices – and as soon as the war was over, the old pattern returned.

Some metal manufacturing did come about, like nails and iron piping for sewage. But it was still not much. That the production of iron and steel products increased from about 90 000 tonnes in 1929 to about 150 000 tonnes in 1940, and that some processed and semi-processed iron products were even exported to other South American countries,[48] was hailed as progress, but was really meaningless. The 150 000 tonnes were still only roughly half of imports, and the total of imports and local production, of about 450 000 tonnes in 1940, spoke very poorly for a country with then about 40 million people.

Real change – in terms of mining and metal-making – came only in the 1930s, with the establishment of the Belgo Mineira steel works, and at the end of World War II, when the Volta Redonda steel plant of CSN (Companhia Siderurgica Nacional) began operations in 1946 – a present of sorts from the Americans to Brazil for choosing the Allied side in the war.

The steel industry – and iron ore

The Portuguese fully expected to find silver in Brazil and, after it became known how much silver there was in Mexico and Peru – especially in Potosí – they were certain that Brazil would also yield untold quantities.

And they did find silvery outcroppings and, indeed, entire mountains of silver, only the silver was not silver but iron ore. Many of the iron ore deposits in Minas Gerais were – and still are – so high grade that fresh fractures glitter silvery.

The first *bandeirantes* were no doubt the discoverers of many iron ore deposits.[49] And, curiously enough, the first iron melting shop, or iron forge (smithy) was set up probably in about 1580 or so in a small township called Santo Amaro, on the banks of a small rivulet that ran – and still runs – into the Pinheiros river. Santo Amaro is today one of the most populous parts of São Paulo.[50]

Eschwege wrote, in about 1815, that the ruins of the old shops could still be seen, including the remnants of what must have been the office and a cooling tank.

But the ore in Santo Amaro only ran 25–40% iron, and the people who worked it – whoever they may have been as no names have come to posterity – were not very successful with their production of iron. But then high grade ore was found in Araçoiaba, just outside the modern city of Sorocaba, in the state of São Paulo. And there one Afonso Sardinha built, in 1590, another small iron forge which he called the Real Fabrica de Ferro São João de Ipanema, in short Ipanema. This has gone into the history books as the first 'real' iron and steel works of Brazil.

Sardinha actually built two furnaces; one of them he gave as a gift to the then governor of São Paulo when, in 1600, the latter visited the works.

Production stopped in 1629, probably simply for reasons of politics. Crown law generally forbade the production of industrial goods in the colonies, preferring such goods to be imported from the motherland. Also, production of only the simplest essentials such as nails and horseshoes made sense in such a sparsely populated hinterland.

All work halted until about 1761 when a new yet much smaller forge was built, but then it stopped again, essentially for lack of coal[51] – weird as this sounds for a country and a region that was then still densely forested, the raw material for charcoal.

In about 1801, a mulatto by the name of João Manso[52] (who had read about iron-making in a book and as a consequence considered himself an expert in the matter) obtained a permit to make iron. He apparently got some help from one Andrade, the local inspector of mines, and together they put up a 'high' (blast) furnace of bricks. They even installed a large bellows, and then invited the local dignitaries and their neighbours to the inauguration, and to the startup. But though reasonable proportions of coal and iron ore were fed into the furnace, and though the leather bellows were worked assiduously, blasting air into the furnace, no iron came out. Manso and Andrade had to flee from the anger of their guests.[53]

The early years of the nineteenth century saw more iron-making activities. In Goiás, a Paulista made a few scissors and knives in 1802, and in Antonio

Pereira, in Minas Gerais, a local slave whose name they forgot to register, made some iron.

It cannot, in any case, have been very difficult to make iron, with the high grade haematite ore found in many parts of Minas Gerais. But it is easy enough to say that now, with the accumulated knowledge of almost 200 years: the reality of the backlands of Minas Gerais in the early part of the nineteenth century may well have been very different.

The years 1809–10 suddenly saw feverish activity. One Manuel Ferreira da Câmara, apparently the administrator of the Diamantina district in Minas Gerais, decided to set up a government owned steel works at Morro do Pilar. His plan was to build three blast furnaces and twelve refining furnaces, so that all of Brazil's iron and steel needs could be satisfied – with a surplus for export.

But the government also had other, somewhat parallel plans. Dom Rodrigo Antonio de Souza Coutinho, the first count of Linhares – who then ran Brazil for the crown – had appointed, in 1809, one Friedrich Ludwig Wilhelm von Varnhagen, a German officer who had earned merits and thanks by fighting Napoleon I on the Portuguese side, to proceed to Sorocaba to inspect the mines and to determine what it would take to set up an iron and steel works.

Varnhagen suggested setting up one blast furnace and three refining furnaces; 150 000 cruzados capital was raised, and the government supplied 100 slaves as well as a mission of Swedish specialists.[54]

History has lost trace of why such a mission had to be called in but the old texts speak of 'friendships' and 'contacts' and 'obligations'. Varnhagen was pushed aside.

The mission that arrived in 1813 consisted of one Carl Gustav Hedberg, in charge, and 13 iron-makers and miners.[55] The trouble, however, was that Hedberg was really a bankrupt foundry operator and his companions were not really experts in anything. One was a carpenter, another simply Hedberg's servant; there were also a tailor, a wood cutter, a shoemaker, a cook, a simple farmer, a stone mason's apprentice, and so forth.[56] The results reflected the make-up of the mission.

Meanwhile Baron Wilhelm Ludwig von Eschwege had decided to pursue his own ambition and to set up his own pig iron works. And he was successful. In December of 1812, Eschwege was the first to produce pig iron on an industrial scale at his plant, Fabrica de Ferro Patriotica, at Congonhas do Campo. Production continued until 1822 when, after Eschwege's departure from Brazil, operations could simply not be continued.

Manuel Ferreira da Câmara began operations at his plant, Real Fabrica de Ferro do Morro do Pilar, in August of 1814 – and fell woefully short of the 40 000 arrobas (about 600 tonnes) a year he had promised. He only got some 395 arrobas (about 6 tonnes) in 1815, and achieved a maximum of 2536 arrobas (about 38 tonnes) in 1820. And Câmara only succeeded after Eschwege had helped out, giving him a master-smith, a German named Schoenwolf.[57]

In 1820, Câmara asked for and obtained two additional German specialists, Hermann and Hans Heinrich Utsch. But they were not metallurgists, rather, apparently, simple labourers with some experience in iron-making.

Work was apparently stopped in 1821 – simply because they could not produce usable iron. By 1831, Câmara's plant was auctioned off.

And what of the Real Fabrica de Ferro São João de Ipanema? For a while, it was a real failure. Hedberg was branded a fraud and Varnhagen was rehabilitated. Construction costs turned out to have been 45 times the original estimate (and budget).[58] The Swedes withdrew in 1815 and Varnhagen was put in charge. He rehabilitated the plant, brought about better production – and then left, in 1821, to return to Europe.[59]

But, curiously enough, it was not the end of the Ipanema plant. In fact, it survived another 70 years. After Varnhagen left, the administration was given, successively, to Captain Rufino J Felizardo, Colonel João F Perêa, Major João Bloem, Colonel Antonio de Melo, Major Dr Joaquim de Oliveira, General Ricardo Gomes, Francisco Raposo (counsellor) and Major João de Lima.[60] Almost all were, judging by their titles, military men, on active duty or retired, and with unknown qualifications. But, though the plant continued to operate, its decline had already begun with Varnhagen's departure.[61]

By 1860, the plant was closed, only to be reopened in 1865 because of the war with Paraguay which caused demand for quick and major steel production.

In 1878, the then owner, the Ministry of War (obviously on behalf of the government of Brazil), transferred the assets to the Ministry of Agriculture 'as the government had resolved to increase production to 20 tonnes/day of pig iron and 1 tonne/day of steel.'[62] Two new blast furnaces were to be built, plus a Bessemer converter. And it was in 1888 that the plant finally made money – 1 conto de reis. That year they produced 600 tonnes of pig iron.[63] And even the quality was good as the iron was used in the foundries of the railroad company to make brake shoes and wheels.

But, as fate has it – and we really lack better explanations – in 1891 the plant reverted to the Ministry of War, and in 1895 it was closed for good. Quite some time later, in 1930, the remnants were converted into a plant to make phosphatic fertilizers.[64]

Today only ruins remain. But Ipanema, the steel-making complex outside Sorocaba, was certainly the longest lasting iron- and steel-making facility in Brazil – and for this reason deserved more than passing mention here.

When all is said, it was not much. No big steel industry was set up, no meaningful 'industrialization' took place. Part of the blame must rest with the stubborn pursuit of something unworkable – the dream of making iron out of high titanium and high phosphorous-containing iron ores, those of the Morro de Araçoiaba.

And this in a country where in the twentieth century huge quantities of almost pure haematite ore would be discovered and exploited. Was it perhaps too much blind trust in foreign know-how, such as that of the Swedes, Varnhagen and Eschwege? And was it complacency, or ignorance, which stopped the Brazilians pursuing other ore occurrences?

The ore was there, all over a large quadrangle, or square, in Minas Gerais, with today's city of Belo Horizonte as its centre. But the time was simply not right.

However, innumerable small foundries, iron furnaces and forges sprang up throughout the nineteenth century all over Brazil, but especially in Minas Gerais.[65]

Two additional pig iron ventures should be mentioned.

The first was Usina Esperança – which in time came to be called Queiroz Junior Ltda and as such, after many modifications, it is still known today – was set up in 1888. This was finally a plant located right where the best ore was – at Itabirito – and in the midst of dense jungle. It also became the first iron plant to use charcoal on a large scale.[66]

The second plant was in Miguel Burnier, also in Minas Gerais.

By 1914, when World War I began, these two plants, each producing but a few tonnes of pig iron a day, were really the only iron-makers of Brazil.[67]

Finally, in 1917, a major step forward came with the founding of Companhia Siderurgica Minera by Cristiano França and Amaro Lanari.[68] Their plant began operations in 1920, in Sabará, Minas Gerais, outside of Belo Horizonte. Later the same year it became Companhia Siderurgica Belgo-Mineira; it is today still a major steel producer in the country.

Lest the inclusion of the word 'Belgo' puzzles, let me explain. Technology from Belgium and Luxemburg was used, and the Belgian Arbed group became a partner. The capacity of the original plant was 25 tonnes of iron per day.

A Siemens-Martin steel plant began operations in 1925, as did several rolling mills. A second blast furnace was started in 1928. A second Siemens-Martin converter began in 1927, the third in 1931. Belgo-Mineira is today probably still the largest charcoal operated steel works in the world.

Another milestone in the history of Brazil's iron and steel industry came in 1930. It was then that a National Steel Commission was set up to study what the country should do – continue to export iron ore and import steel, or set up its own steel industry. Studies and deliberations dragged on until eventually consensus was reached that the country should have a large and fully integrated steel-making complex.[69] By then it was 1940, and World War II had begun.

Remembering severe shortages of industrial goods, and also of raw materials during World War I, fears now arose that a protracted – and global – war would again affect Brazil badly. This is how – and when – Companhia

Siderurgica Nacional (CSN) came about (1941). It began operations in 1946.

In the political pulling and pushing over where to locate the plant, Rio de Janeiro, then still seat of the federal government, won out over São Paulo and Minas Gerais. The location at Volta Redonda, a town somewhat south of Rio de Janeiro, was thought to be about equidistant from the mines in Minas Gerais, the port of Rio de Janeiro – for importing coal from abroad or shipment of domestic coal up the coast from the southern states – and the principal markets for the steel, Rio de Janeiro and São Paulo.

Interesting stories are still bandied about concerning the origins of CSN. It was, supposedly, a gift from the US government, a bribe to bring Brazil into the Allied camp during the war. And, indeed, the original equipment was all of US origin, as was the money that paid for the plant – loans from the Export–Import Bank.

Secret documents about a deal between the United States and the government of the then Brazilian president (and dictator) Getulio Vargas have not come to light, and there may simply not have been a deal, just a number of coincidences – that Brazil finally wanted a steel industry, that the United States wanted air bases in north-eastern Brazil, and that Brazil did declare itself on the Allied side.[70]

And so Brazil got its big steel plant in 1946. CSN is still operating today, in the early 1990s, but, of course, is much larger now. Its principal problem today is a debt load to foreign lenders and domestic suppliers, of staggering proportions.

When, in 1974, the Brazilian government set up Siderbrás, the holding company for the state owned steel works, CSN became an early member of the behemoth. By then, another big state owned steel plant had been built near the port of Santos: Companhia Siderurgica Paulista (COSIPA), founded in 1953.

By the late 1980s, the Siderbrás group had grown to be the largest steel group in the country, with CSN at Volta Redonda in the state of Rio de Janeiro; COSIPA at Santos in the state of São Paulo; USIMINAS at Ipatinga and AÇOMINAS at Ouro Branco, both in Minas Gerais; and CST (Companhia Siderurgica de Tubarão) at Vitoria in Espirito Santo; plus a number of smaller units spread all over the country.

In addition, quite a few private steel producers were operating, also spread across Brazil, among them Villares, Acessita, Mannesmann, Belgo-Mineira and the Gerdau group.

In 1990, President Fernando Collor de Mello took the decision to extinguish Siderbrás, the holding company, and gradually to sell off the state owned steel mills to private investors, both domestic and foreign. USIMINAS was the first to be auctioned off, late in 1991, and CST and some of the smaller works were to follow in 1992.

Brazil's steel production for the first half of 1991 reached 10 960 000

Table 12.3. Production of pig iron based on the use of charcoal (million tonnes)

Year	Integrated steel plants	Independent producers	Total	Pig iron exports
1980	2.5	2.4	4.9	0.85
1985	3.0	3.8	6.8	2.5
1989	3.6	6.1	9.7	2.9
1990	2.9	5.6	8.5	3.4
1991	2.9	4.5	7.4	2.56

tonnes, some 8.4% above the production during the first half of 1990. With this tonnage, Brazil accounted for 57% of total Latin American production.[71]

Brazil has also had, since the early 1970s, a rather substantial pig iron industry, innumerable plants and furnaces that produce principally foundry grade pig iron (low in phosphorous and low in manganese) by using charcoal as fuel. Much of the iron is exported.

The charcoal comes principally from cutting and processing virgin (jungle-type) forests in ever remoter areas of Minas Gerais, Espirito Santo, Rio de Janeiro, Bahia and other states.[72] This procedure is, naturally, heavily criticized, not only abroad but also inside Brazil as depleting the natural resources of the country. But at the same time, an ever larger part of the charcoal comes today from 'reforestation farms' where, from about the mid-1970s, legal requirements (often, unfortunately, laxly administered) forced charcoal consumers to replant tree by tree the 'natural resources taken out', to thus make charcoal a renewable resource.

Commendable as the effort is, the process has been slow and it will still take a few years before charcoal from reforestation farms will, in terms of quantity, come to equal the coal produced from virgin forests. By 1989, it was only at 29%.[73]

Production of pig iron based on the use of charcoal has reached the levels shown in Table 12.3. In addition, the integrated steel works produced a further 12–15 million tonnes of pig iron each year based on coke.

Steel and pig iron production is dependent on the availability of iron ore – and iron ore is a resource that Brazil has more than enough of. Much of the area around Belo Horizonte, in the state of Minas Gerais, consists of iron ore, and innumerable mines produce what is basically very high grade and very pure haematite ore.

It would lead too far to give here more than a bare outline of the industry, but two names stand out: that of the Companhia Vale do Rio Doce (CVRD), a true giant among world giants in iron ore, and Carajás. This latter complex, owned by CVRD, will be discussed below.

The Vale company – as CVRD is generally referred to in Brazil – came about in 1942, as part of the country's move into big-time steel-making.

The origins of the company go back further, to the beginning of the century, to the original Companhia Estrada de Ferro Vitória-Minas, the old railroad built in 1901 to transport ore to the port of Vitória. They went back also to the major surveying and mapping of the Itabira iron and manganese ore deposits that took place about 1907. Some 4 billion tonnes of 65% iron plus ore were identified.

British interest, through the Brazilian Hematite Syndicate played a major role, as did the Itabira Iron Ore Company Ltd. The latter belonged to US, British and German interests and was run by an American engineer (and tycoon), Percival Farquhar.

Farquhar had been involved in a number of mining projects in Brazil and in 1919 had proposed a master solution to the nagging iron ore and steel problem.

His suggestion was that his company, which owned large ore deposits in the Itabira range – and apparently also controlled the rail link to Vitória – would rebuild the railroad (presumably to make it suitable for large scale ore shipments) and build a new port to export 5 million tonnes of ore a year and to receive the foreign coking coal he proposed to bring to Brazil in the same ships that would take the ore.

To get the project approved, he also offered to set up in the Rio Doce valley an integrated iron and steel works with a capacity of 150 000 tonnes a year.

The Itabira Iron Ore Company Ltd would with one stroke have obtained a monopoly hold on Brazil's iron ore exports and its steel-making capacity.

President Epitácio Pessoa approved and a contract was signed in 1920. But the project never got off the ground. Minas Gerais politicians (notably among them Artur Bernardis, then governor of Minas, 1918–22, and later president of Brazil, 1922–6) objected, and as foreign financial backing was subject to a guaranty from the state of Minas Gerais, which was not forthcoming, the 1920s passed with no action. The 1929 crash and the subsequent worldwide depression delayed the project further and by the late 1930s, sentiment in Brazil had turned nationalistic; CSN, at Volta Redonda, was then the solution to the steel problem.

As part of the 1942 agreements in Washington, or concurrent with them, CVRD came about, ending up as owner of the Itabira Iron Ore Company, its deposits, and also the port and railroad installations.[74]

CVRD grew, over the years, and eventually achieved pre-eminence in the world. A few statistical details for iron ore exports in the period 1950–91 are given in Table 12.4.

In 1991, the country produced a total of 151 million tonnes of iron ore. CVRD was responsible for about 62% of the total; MBR/Caemi, the second

Table 12.4. Iron ore exports, 1950–91

Year	Tonnes
1950	900 000
1960	5 200 000
1970	28 000 000
1973	37 500 000
1981	45 900 000
1989	113 700 000
1990	112 500 000
1991	120 000 000 (estimated)

largest group, accounted for about 12%. Exports in 1989 came to 113.7 million tonnes; the two largest producers accounted for 92% of the total.

CVRD is today involved not only in mining iron and manganese ore, bauxite and phosphate rock, but also in making iron pellets, aluminium metal, gold mining, other metals and ores, as well as cellulose.

Scheelite and wolframite

These ores are minor in the context of Brazil's overall mining picture, but they deserve brief mention.

Wolframite was apparently mined somewhere in the state of Santa Catarina at the beginning of the century and again in the state of São Paulo during World War II, but at both times the quantities were negligible and soon forgotten.[75]

But in 1943, a single large scheelite deposit was discovered at the farm of Tomáz Salustino, near the town of Currais Novos in the state of Rio Grande do Norte. For a while, apparently until about 1953, the deposit was worked by *garimpeiros* before more orderly operations could be started. Four or five companies continue today to work the ore body that slants at about 15 degrees into the ground away from Salustino's farm.[76]

And in 1980, alluvial wolframite deposits became known in the Amazon region. It was, precisely, somewhat to the south of the town of São Felix do Xingú, just off the Xingú river, in the state of Pará, that *garimpeiros* discovered fine-grained black sand (or at least it appeared to be a kind of sand) in creek beds and dry gulches.

The *garimpeiros* at first thought the material was cassiterite – that was what they were looking for, besides gold – of which much had been found, and is still being found all over the Amazon basin. But analyses revealed the ore to be wolframite. Soon they also found outcroppings, and then shallow veins.

It took the trading and, in general, the mining community a while to catch on to this availability of a new ore – and the *garimpeiro* population for a while grew in leaps and bounds.

These were also the years – the early 1980s – when, at least for a while, tungsten prices were exceptionally high. Only this explains why primitive manual processing made financial sense, and even flying the resulting concentrates out, to São Felix do Xingú, in single engine aeroplanes, from otherwise almost inaccessible locations.[77]

The wildness, lawlessness and primitive conditions of the *garimpeiro* camps gave insight into what earlier gold and other metal rushes in remote locations must have been like. During the dry season, water had to be carried by mule to the camps. All food took the same route. Women were not tolerated, also no liquor. Nightly entertainment was pot-shooting at soda cans, across log fires, and into the dark wilderness beyond. Violent crimes were rampant, and scores were settled at once. The law authorities themselves came rarely and solved nothing.

Those outsiders who braved the hardships of travelling there, the dangers of using planes already discarded elsewhere, the shooting of rapids up the Xingú, the heat, the insects, the simple and often crude food, and the lawlessness of the backlands, could experience real adventures.[78]

The *garimpeiros* simply worked what they found, and bothered little with who owned the deposits. As it turned out, CVRD was the owner. In time, lower prices drove the *garimpeiros* away again and by 1990, the *garimpo* Pedra Preta is once more awaiting opening up and industrial exploitation.

The story of these *garimpos* near the Xingú river needed telling as their existence was unsuspected until about 1980. And there the wild prospectors suddenly discovered a new mineral (wolframite), many miles from anywhere, right in the wilderness. What else is there still around, in the vastness of the Amazon lowlands, between the borders with Bolivia, Peru and Colombia to the west, the borders with Colombia and Venezuela to the north, and the Atlantic Ocean to the east? Untold riches, vast ore reserves, major deposits waiting to be discovered?

A very substantial tin industry

By 1989, Brazil was producing 51 000 tonnes of tin metal, roughly 20% more than the year before, roughly double the quantity produced in 1986, and five times the production of 1980. If one goes back as far as 1967, production then was a paltry 1800 tonnes – and back in the early 1950s, production of tin (mining) had not even existed.

The ore had, of course, been there all along. Only nobody had known it. Much of the ore may well be of Bolivian origin, washed down before the dawn of time from the Andes, washed out of the mountains and swept down into the Amazon plain.

In time, it was found in Rondônia and Mato Grosso, at the foothills of the Andes, and in Roraima, and in Pará, near the Xingú river. And there may well be other places where tin has not yet been found, where it has simply not yet been looked for. There are still many areas in the north-western and northern parts of Brazil where little penetration has taken place, and if so, only by rubber tappers and gold prospectors. And gold prospectors would generally throw black dust out of the pan, especially when they are deep in the jungle and far away from civilization so that transporting ore would make no sense – even if the ore could be crudely upgraded to a 60–65% concentrate worth maybe US$4000 a tonne.

Notwithstanding the foregoing, it was indeed a rubber tapper, or *seringueiro* (latex gatherer), who did find tin in Rondônia, in about 1956/7. One day, during his wanderings, this particular rubber tapper crossed a ravine, somewhere some 100 or so kilometres south of Porto Velho, the capital of Rondônia. He saw outcroppings of unusually black rock particles, standing out among much lighter-coloured surroundings.

The rubber tapper was not familiar with mining, and he was not searching for minerals. His only tools, knives and a small axe, were inadequate for breaking up the rock. But he collected a sample quantity.

It was the sample of ore which eventually appeared in Porto Velho, the next time the rubber gatherer brought his load of latex balls to the owner of the land. The rubber tapper had noticed that the black particles were substantially heavier than the other pieces of rock and gravel in the ravine.

The sample turned out to be almost pure cassiterite. It was analysed in São Paulo, as no laboratory existed as yet in Rondônia.

It had taken a few months for the results to get back, and then some more time was lost before somebody was found who could interpret the assay certificate. And even more time passed before it was possible to get an idea of the value of the ore which, it turned out, was many, many dollars per tonne.

Until that moment, other rubber tappers, all simple men from the jungle, had chided the discoverer of the ore for wasting time, running after black dirt and for hoping to get meaningful answers out of those 'fanciful southerners', as the people of São Paulo were called. But all that changed after the results reached Porto Velho and became known, and especially after Joaquím Pereira da Rocha, the owner of the land where the *seringueiro* worked, had told the latter to go back into the jungle, to take a second and a third hard look, and to bring out more samples, from up and down the ravine for as much as half a mile each way. The rubber tapper had even to blow out a few holes with dynamite, of which he was given a few sticks.

In time, Joaquím Pereira da Rocha would become known as the man who discovered tin in Rondônia, although he was just the owner of the land where the itinerant rubber tapper found it.

The rubber tapper not only found outcroppings, but also fine black sand – again unusual, among so much light-coloured sand – at the bottom of the creek that flowed through the ravine. He also packed samples of this sand and then patiently lugged them to the river for transport to Porto Velho.

The second trip yielded not just a few pounds of material, but a whole canoeful, close to 200 pounds. The sand, too, was significant: it turned out to contain high grade cassiterite tin ore. And Joaquím knew he was on to something really new and big.

Joaquím eventually said something about 'his' find to a friend, who in turn mentioned it to another friend and, in no time, the story made the rounds. But the first problem was that only the rubber tapper knew exactly where the ravine was. Joaquím had a vague notion; the area was, after all, his and he had been there often. But he was not as familiar with every creek and ravine as the rubber tapper. Others only knew generally where Joaquím's concession was located.

Strangers never knew where a latex gatherer worked, and the *seringueiro* himself would not tell where his tin cans hung, and where his rounds were, so that others would not sneak up on him and steal the fruits of his labour. Occasionally a stranger did anyway, when he accidentally came upon a tree with pots, or recognized a gatherer's circuit, and then the owner would have to fend for what was his. They were thus all armed and many a dispute was settled quietly (or loudly, depending on whether a knife or a gun was used), without the outside world becoming privy to the details.

The rubber tapper, who would soon disappear completely from sight, was illiterate and thus unable to fill in a map. But Joaquím needed to know the exact location of the ravine – and he had no choice but to go into the jungle himself.

They – Joaquím and the rubber tapper – now took better equipment and Joaquím was determined, and prepared, to search more professionally – not that he had much of a notion what that meant – and to take further samples.

And so they did. Once they got back, others started to ask questions more insistently. But the two only told how they had found the ore, what the outcroppings looked like, what the deposits in the creeks looked like, and how the black ore was simply heavier and blacker than anything else nearby. They assiduously avoided giving even a hint of where they had found the ore. But they did say that the creeks and ravines and dry riverbeds of the region ought to be full of the black stuff. Joaquím admitted making these comments somewhat off the top of his head, without really knowing this to be true, but some 20 or so years later when Joaquím Pereira da Rocha was long since dead, some people would say that he had been absolutely right, because quite a few ore deposits, and major ones, were in time found on what had been his concession, and in the adjacent areas, and these were then worked on a large and industrial scale.

Prospectors (*garimpeiros*) mining cassiterite deposits in the Amazon (by Barboza Leite). Source: reproduced by permission of Revista Brasileira de Geografia and of the Fundacao Instituto Brasileiro de Geografia e Estatistica – IBGE.

The immediate results of Joaquím's comments and the positive results of his samplings were that other rubber tappers began to look for cassiterite; in time they also began to find it. And over the years the latex gathering declined, as searching for tin ore became more profitable and attracted more and more people.

Unlike rubber gathering, where one man can take care of a large area tapping the usually dispersed trees, mining for tin ore was not work for just one man. Once a deposit was discovered, there was room for many, and often thousands would gather to exploit one 'mine'. The site then quickly became a frontier settlement, with a crude form of administration and justice. Might often substituted for right; weakness and sickness were often fatal.

Joaquím Pereira da Rocha quickly found out that his secrecy was of only limited value; in time he was successfully followed and others began to work in the same ravine. A few shots were even exchanged. But Joaquím accepted that his legal rights were non-existent and that his survival depended on a successful working arrangement with his neighbours. As the years passed, he came to be regarded as a legend, a *primus inter pares*, and as a leader, not only because he had discovered the ravine but also because he jealously guarded – and worked – the higher grade claims himself.

The wild prospectors now moved in, most of them former rubber tappers, but also adventurers and drifters, as well as unemployed men who dared to face the wilderness. Not all succeeded; quite a few went back. And some succumbed to illness or violence.

It took only three years before the original ravine was depleted. Most of the finer material was left behind. Many of the poorer grade deposits were ignored, and much of the richer material at the bottom of the creek was just muddied up, dispersed, and covered with layers of dirt and chips of rock from the side of the ravine, as the prospectors went first after the rich outcroppings on the sides. By the time work on the sides had become non-rewarding, the bottom of the ravine was covered, in fact quite well buried, and whatever ore was at the bottom of the creek was lost.

As the diggings became poorer, even if only for a short period, some of the prospectors began to drift away, every attentive to the rumours of other, richer strikes nearby. One of the richer strikes eventually became the Santa Barbara mine, worked by a firm called Mibrasa, today part of Cia de Mineração Jacundá SA, belonging to Canada's Brascan group.

This mine had originally also been no more than a dry riverbed where, in the mid-1960s, several hundred *garimpeiros* had established themselves, to take out the rich surface pickings.

It was at another Mibrasa mine that an event occurred which has been told and retold many times and which has by now become a legend.

It was about a man called Cassio de Souza Mello, an engineer of the Mibrasa company which held the exploration and exploitation permits for the deposit. In 1968, Cassio arrived at the mine site to carry out a survey. But he was kidnapped, or arrested, by the *garimpeiros*, as likely to endanger their livelihood.

They held a quick trial and, in the manner of frontier justice, found him guilty without much accusation nor defence. He was condemned to death. Cassio was taken to the main square – simply the centre of the small *garimpeiro* settlement at the mine site – and he then began to talk. Just like that, simply talking, softly, and with a smile. Do not ever show fear to these devils, he would later say when speaking about the incident. And some 15 or 18 hours later, the next noon, he was still talking and still standing and still smiling. He had talked through the entire night.

The prospectors who the previous evening had wanted to do away with him, had eventually come to admire his stubbornness, and his stamina, and his smile; they finally declared him their friend and their spokesman, and carried him on their shoulders to a few weakly shouted vivas, as by then they were all exhausted.

Cassio had not defrauded them, but had really acted as their friend in the

Table 12.5. Smelters converting cassiterite into tin metal, 1990

Smelter	Tonnes
Paranapanema	18 500
Cesbra	4 500
Corumbatai	3 000
Best	3 000
Minoeste	2 000
Bera	1 400
Mequimbras	600
Fluminense	500
SNA	3 000 (?)
SNC	500 (?)
Total	37 000

dispute with the company. And over the years the incident and its outcome became a legend; Cassio became a legend in all of Rondônia, and also to the thousands of prospectors, many of whom had never met him, yet admired him and considered him their friend.

The ore – in the beginning handpicked and hand-concentrated by meticulous selection and elimination of any material that was not strictly pitch black – came out of the hinterland to Porto Velho, and from there found its way by vessel to Manaus and on to Belém. From there it was transported by vessel to Rio de Janeiro, and on to the then only existing tin smelter in Brazil at Volta Redonda (where the tin would go right into CSN, the only tinplate mill in the country). Ore buyers at Belém and in the south picked the movement up from the statistics and traced the ore back to Porto Velho. But it was the owners of the smelter, which had been built quite a few years earlier, in 1952, to process imported ore and concentrates from Malaysia and Bolivia, who sent an ore buyer to Porto Velho as early as 1962.[79]

It was this ore buyer who asked questions of the *seringueiros* and the traders and who began to map what was then the territory of Rondônia to discover where the ores were coming from. A few years later, geologists and engineers followed, exploration and exploitation claims were filed and more orderly mining began.[80]

Brazil's rise to prominence as a tin producer has been spectacular. In 1967, the country was still in eleventh place among tin-producing countries; by 1976 it was in eighth place, by 1985 in second, and in 1987 Brazil was the largest tin-producing country in the world, with a total production of 54 000 tonnes in 1989. Indonesia and Malaysia are distant runners-up.

Back in 1967, Malaysia had been the top producer, with 73 000 tonnes, followed by Bolivia, with 27 400 tonnes.[81]

Table 12.6. Brazil's share of world tin production, 1978–88 (tonnes)

	Brazil's production	World production	World consumption	Brazil's share of world production (%)
1978	6 500	235 800	231 800	2.4
1988	42 800	203 800	240 600	21.0

In 1990, a number of smelters were operating in Brazil, converting cassiterite into tin metal. These are shown in Table 12.5. Total estimated production for the year was 37 000 tonnes,[82] which ties in more or less with Brazil's estimated export tonnage of 30 000 and expected domestic consumption of about 7500 tonnes. It is unclear how much was smuggled out of the country (in 1989, the smuggled quantity reached about 10 000 tonnes[83]).

Brazil's production increases over the last 10 years – almost eightfold in the 1980s – have caused substantial changes in international markets as is shown by Table 12.6. Without Brazil's dramatic production increases – which made up for similarly dramatic production declines in other countries – the world would have seen severe supply shortages.[84]

Carajás and Serra Pelada

The big mining stories of the 1980s were Carajás and Serra Pelada, the first thought of principally as an iron ore deposit, the second as an open pit gold mine. In a very general way, both may be part of the same mineralized zone, being no more than some 60 kilometres distant from each other and both surface deposits.

The Serra dos Carajás is located some 550 kilometres straight south from Belém. It is a mining complex of staggering proportions. The reserves of iron ore are said to be upward of 18 billion tonnes of highgrade (above 66% iron) haematite ore – enough for almost 500 years of production at the current rate of about 35 million tonnes of yearly output.

Only a small part of the complex has been opened up – N4E; it is an open pit operation. The ore goes by giant trains some 800 kilometres to the port of Ponta de Madeira in São Luis, in the state of Maranhão.

Like so much in mining in Brazil, this deposit also became known rather late. One or other prospector did previously reach the area but nothing of significance was noted anywhere until, as late as 1966, a subsidiary of Union Carbide discovered manganese ore in the area. Then it was United States Steel that located and identified major iron ore deposits in the area.

Companhia Vale do Rio Doce next became interested and proposed a joint venture to US Steel; this led, in 1970, to the formation of Amazônia Mineração

SA (Amza), of which CVRD controlled 51%; US Steel held 49%. The extent of the deposits was checked out in the early 1970s, and by 1976 the building of a railroad to the nearest port was authorized and begun.

In 1977, US Steel sold out – its 49% went for $55 million. And in 1981, Amza was merged into CVRD.

The mining complex is known to include large bauxite and manganese deposits, as well as copper, gold, nickel and tin deposits, beside the huge iron ore mines.

The railroad – meaning exports could be made possible – was the key to production, and when it was inaugurated in 1985 production and exports began. By 1990 this reached the level of 35 million tonnes per year.

The discovery of the iron ore deposits is usually associated with the early use of helicopters by US Steel, in 1967, to explore, or map, the area south of Altamira-Tucuruí, west of Marabá and Xamboiá, and east of São Felix do Xingú, or between the Araguaia and Xingú rivers. Helicopters were good for exploration as they fly neither very fast nor very high. And this is how bare hilltops were seen and then looked at. They were bare because they consisted of iron ore, and high grade ore at that, at 66% iron almost pure, and thus the hills had little vegetation on top.

Somewhat earlier, in the 1950s, aerophotographic mapping had already shown the bare spots, but the occurrences had been interpreted wrongly, as being of calciferous sandstone.

Why did US Steel sell out to CVRD? Why did it not keep the deposits it had discovered? Or keep at least its 49% share? It appears that when US Steel filed the claims for the area with DNPM, the Brazilian National Mineral Production Department and the agency that registers and grants such requests, the officials of the agency, realizing from the size of the claims that US Steel had maybe found something big, alerted the government company CVRD to send a geologist to investigate. Once the extent of the deposits became clear, the government itself was alerted, and negotiations with US Steel were begun that ended, just a few years later, first with the joint venture in 1970, and in 1977 with US Steel bowing out.[85]

Carajás was by 1991 the largest iron ore mine in the world, and the Carajás complex of mines is likely to turn into one of the most diversified and richest mineralized zones of the world.

When CVRD went to a production level of 35 million tonnes per year, it probably reached a temporary plateau. Total iron ore production in Brazil reached 154 million tonnes in 1989, some 46 million tonnes more than Australia, the second largest producer in the world. And of the 154 million tonnes, 74%, or 113.7 million tonnes, were exported.[86]

In February of 1980, during the carnival period as legend would eventually have it, gold was found in the Serra Pelada. For once it was not *garimpeiros* who found it but the peons of a farmer who had tried his hand at cattle ranching but had finally almost run out of money and was trying to sell his 7 million acre ranch. The gold find was thus most welcome.[87]

By today, some 13 years later, nobody remembers any more what the countryside looked like. But it was probably the same kind of hilly jungle still found over most of the states of Pará and Goiás – few tall trees, much underbrush, a few creeks, many dry creeks and riverbeds, a few big rivers. In other words, rather inhospitable nature, bone dry in the local winter (July–October), quite wet the remainder of the year.

Some people say the first gold was found in a dry riverbed. It was certainly all alluvial, with much mud, or dust, depending on the time of the year, and the gold in the form of dust and sand in fine grains, with a few nuggets. The largest nugget they ever found at Serra Pelada weighed something like 75 kilograms, and it netted the finder a cool 1 million dollars.

The farmer – Genesio Ferreira da Silva – and his peons could not keep the secret and in no time 30, then 50, then hundreds and finally thousands of prospectors came, to dig and to pan for gold.

The total of those that rushed to the gold diggings of Serra Pelada supposedly eventually reached 100 000.

It must have been the largest gold mine ever in the history of Brazil, and maybe even in the world. And while 100 000 or so people lived there, it was probably one of the largest cities in the Amazon basin.

But by the winter of 1984, the numbers began to go down again. The best pickings, the rich spots, had been taken out and the quantities of gold found diminished.

Getting to Serra Pelada was, in the early days around 1980, a major undertaking. By April of that year, only two months after the discovery, a crude runway had been built, and small, one engine planes flew in, taking four or five passengers at a time, at a high price. The operator had a deal with Genesio, the owner of the land: between them they had built the runway, and whenever anybody else wanted to land, they just blocked the runway, with a cart, or a log or two. These kind of shenanigans did not last too long, but while they lasted, the service was extremely lucrative for the operator and for Senhor Genesio.

The other way to get to Serra Pelada was, literally, on foot. People would hire a taxi and a driver and have themselves – often several together – driven, obviously also at a high price but still cheaper than by plane, some 100 kilometres over what was most of the time just a muddy path or track, to where that track ended. From there they then had to walk another 10 or 12 miles through the jungle.

Soon enough it was a clearly marked path of trampled-down vegetation with a lot of litter thrown along the way. The walk could take anywhere from 4 to 10 hours, depending on the weather and the quantity of gear and supplies a man was carrying.

By then, April 1980, Serra Pelada was already a sizeable settlement, a village, or town, because some 15 000 people must have been there.

It was just a filthy dump. It was – said those who saw it then – like the worst kind of slum in Rio de Janeiro. Housing was just ramshackle – huts, lean-tos, tents, tarps and plastic sheets hung like tents. There were very few constructions that could have been called buildings. There was, after all, no way to get anything to the place except by plane or by foot. So housing consisted of what people could carry in, and what they could make with what they found there, tree trunks and branches, leaves, and so on. And they had not been able to dig holes, or had not wanted to, because it was always raining, and what good is a hole in the ground as a home when water collects in it?

It was a town without much of anything. There was nothing that could have been called a street. But there was a square, more or less in the centre of the town, and they called it the 'Square of Lies'. This was because it was there that they gathered in the evenings and where they swapped stories and told each other lies, bigger and bigger ones, of the gold they had found, and what their dreams were, and the women they had had, and what they would do when they struck it rich the next time.

No women there? No! Not one! Women were not allowed, in those early days. They, the men, wanted to avoid prostitution and fighting. They also had a prohibition against liquor. No *aguardiente*! And no gambling!

Genesio had originally established these tough rules. When the federal police arrived, a little later, they kept and enforced them.

There was garbage all over. People urinated and defaecated wherever they pleased. There was no sanitation, no public toilets. When the numbers had been few, toilets appeared unnecessary. But with thousands . . . ?

But the physical aspect of the town was only the first shock to the new arrival. The actual mine, the pit, the Gruta Rica, the rich hole, was an unbelievable sight.

The 15 000 had by April 1980 not just taken on the bottom of the ravine, where the original gold had been found, but also the side walls, all the way up to the rim. They had apparently found gold all over the place, so it mattered little where they worked.

Looking at the pit from the top was like a vision of another world.[88] Mining engineers would have described it as an open pit, but an unknown prospector with a sense of history called it 'Babylonia' where, as legend had it, nameless and numberless slaves toiled to build monuments.

And the name stuck. At this new Babylonia it was greed and hope and ambition that made people work, and they were all slaves to these emotions, from the very first day.

The pit looked like innumerable blocks of earth, like cubes, next to cube-like holes. These were the claims. Each was like the next. All were the same size, 2 metres by 3 metres. Specific individuals had title to a claim. There were then, in about May 1980, some 4000 such claims.

It was, in the beginning, a very small area. But it grew quickly, spreading along the ravine and spilling over the rims.

Each tiny plot would, in time, be duly registered and covered by a title deed. Savagery and wildness did not last long; the authorities, and with them bureaucracy, came soon enough.

At each claim there were one or two men who loosened the earth – coarse gravel and powdery sand which became mud when wet – with picks and shovels. Then they filled the stuff into plastic bags, mostly the woven, strong type.

A bag would hold between 40 and 50 pounds of 'gravel'. The bags would be taken up out of the ravine to the rim beyond the digging area for processing. The men who did the carrying did as many as 40 trips a day, from dawn to dusk. The rule was that each of them was expected to bring up about a tonne of gravel a day.

Parts of the rim were terraced, and the routes up often zigzagged up and sideways and up again. Each part of the way was difficult, and some parts were very dangerous. Once a man had a bag on his shoulder, he would negotiate his way around hundreds of people who seemed to be aimlessly milling around – though each really had his task. He would go around holes and upright blocks, or claims, to the first elevation, from where he would then climb up some crude ladders which were nailed or lashed or maybe even just tied together. He would climb from terrace to terrace, until he reached a more gentle incline on the way up.

All the peons looked about the same. All wore shorts, most had a shirt, or what was left of a shirt, some wore straw hats or bonnets, and a few even plastic helmets. Some just put an empty bag over the head, and the length of the empty bag that fell down the back would protect their neck and shoulders.

Since much of the dirt was wet, the bags were wet and dripping, and the men were also dripping. Malaria, and pneumonia, as well as all kinds of rashes, fungi and so on, were rampant.

At the rim, or near the rim, depending on where a claim holder had his set-up, the bags were emptied into piles, one pile per claim. That was also where the washing took place. The most promising material was panned, the rest sluiced. Water was the most important raw material and they brought it in buckets from nearby creeks. The fancier outfits, or claim holders, brought it in with rubber hoses or crude channels.

Those that hired themselves out as peons, as gravel carriers, doing 40 or so trips per day, up and back and up and back in what was back-breaking work, at fixed pay and food, never got rich. The pay was extremely low. It was just another form of slavery.

Only a few claim holders had real money on them. A lot, maybe most, of the transactions were done on trust. And for food.

A claim owner would have six or eight peons, one or two to shovel gravel into bags at the claim, three or four to carry the bags up, and yet another two or three to work at the washing end. They all worked for a daily fee and food. And if there was no money, they just worked for the food and waited. And were owed the money.

Money was available only when there was gold, and only once the gold had been sold. If a claim did not produce for a while, payments would be really slow. But the system worked. When a claim did not produce anything for a really long time, and the debts mounted, the owner sometimes ran for it. He simply fled. That left the peons unpaid, and the supplier of food and so on unpaid. But it happened rarely. People just had faith, in those early years, that a claim would produce, and it usually did. But only very few struck it really rich.

When a man abandoned his claim, didn't work it, or ran away, it took only 24 hours before the claim was reassigned to somebody else. And it was often not just one man who held a claim; many were owned by two or three, or more, as a group.

Once a peon got paid, and his debts had been deducted, or he had been forced to pay up – loans and supplies and meals and so on cost enormous amounts, including sky-high interest charges and fees – and he still had any money left over, he usually did not know what to do with it. Or where to hide it. For a while, belly pouches were the rage. The merchant who flew in a few hundred sold them in three days and made a fortune. Where else would the peons have kept their money? The huts were not safe, banks did not exist, and to go into the jungle and dig a hole somewhere to hide it too risky.

But most of the peons never got that far. They rarely had enough to hide. They found that just as life in a big city slum was tough and profitless, it was tough and profitless at Serra Pelada too. But at Serra Pelada, they at least had some hope. Each would dream of finding the big nugget, or at least *a* nugget, and then be rich beyond his wildest dreams, and for the rest of his life.

There was no end to the stories of personal tragedy and personal triumph; some struck it rich, but most did not, and over the years, as the grades got poorer, most drifted away again and fewer and fewer stayed.

As they worked themselves deeper into the ground, the sides got steeper and steeper, and they had cave-ins and deaths. And when it rained for two or three months, all work had to be suspended as it became more and more dangerous.

By 1991, only a few thousand were left, reworking the dumps and patiently trying this or that end of the pit.

The total quantity taken out of the pit may have reached 200 tonnes.[89] But this is guess work. Reliable statistics do not exist. Much gold was undoubtedly smuggled out and was never picked up by the statistics. For a number of years, smuggling was very profitable as black market dealers bought the gold paying the black market exchange rate when converting the dollar value of the gold into local currency, or paying in dollars directly, always a desirable currency in Brazil where, in those years, the local currency depreciated by 50–100% per year, and then by 200–300% per year, and for a few years at still higher rates. And he who sold his gold for dollar bills was also not subject to this and that small tax, and also not subject to income tax which he would have had to pay if his gold was sold to a government bank and the sale was duly registered.

CVRD held the underlying claims; all the gold should have been theirs. But once the *garimpeiros* took over, and thousands and tens of thousands worked the pit, their presence made it a social problem. And a political one. CVRD could have secured the ravine only through massive police intervention, something neither the company nor the government was willing to face. The Serra Pelada gold rush was thus allowed to run its course, until the ravine and the hills and the ground were exhausted, and the *garimpeiros* left. But much gold is still there. Indeed, probably as much as half of the gold is still there. The primitive washing and sluicing methods of the *garimpeiros* often made for losses well in excess of 50% of the content of the gravel.[90]

It will still take a few years before all the *garimpeiros* are gone; then CVRD will undoubtedly fence the area off, bring in large and sophisticated equipment and rework the entire area, to recover the rest.[91]

Bauxite, alumina and aluminium

It begins with a clay – one of the most common types of earth. It is called bauxite, and is the same as that found in Jamaica, in the Dominican Republic, in Venezuela, Guayana, Surinam and elsewhere. It generally contains between 40% and 60% aluminium oxide. When it is refined to a white powder it becomes almost pure aluminium oxide, and is then called alumina. And when alumina is dissolved in molten cryolite, and then broken down by passing an electric current through the molten mixture, it separates into aluminium and oxygen.

About 2 pounds of alumina give 1 pound of aluminium, and as bauxite contains 40–60% aluminium oxide, 4 pounds of bauxite give about 1 pound of aluminium.

Brazil is rich in mineral deposits, as we have seen, and it was and remains logical that there is much clay and much bauxite all over the country – as indeed there is.

In the early 1990s, Brazil is a major producer of all three, bauxite, alumina and aluminium.

The country's first venture – adventure, really – into aluminium production occurred in 1938. The desire to get aluminium for the production of aeroplanes – the shadow of the upcoming world war loomed large – led to the formation of Elquisa. It started production in 1945, but almost immediately shut down in 1946, when excess production capacity elsewhere in the world,

Table 12.7. Aluminium production, 1951–90 (tonnes)

Year	Company	Production	Total yearly production
1951	Alcan	400	400
1961	Alcan	9 600	
	CBA	8 270	17 870
1969	Alcan	22 824	
	CBA	20 100	42 924
1971	Alcan	25 129	
	Alcoa	7 900	
	CBA	23 118	56 147
1981	Alcan	87 345	
	Alcoa	88 537	
	CBA	80 536	256 418
1984	Alcan	119 595	
	Alcoa	116 337	
	CBA	137 916	
	Valesul	91 151	464 999
1990	Albras	160 000	
	Alcan	118 000	
	Alcoa	90 000	
	Alumar	255 000	
	CBA	180 000	
	Valesul	92 000	895 000

and lots of scrap worldwide, drove prices down and the small, experimental producers out of the business.

Meanwhile, the metal was not unknown in Brazil. Utensils had been made locally ever since 1913, and in 1915 Alcoa and Alcan had set up sales offices for important aluminium.

The first major step towards large scale production took place in 1941 when the Brazilian Votorantim group formed CBA, Companhia Brasileira de Aluminio. It began production in 1955 – at what was to become a 10 000 tonnes a year operation.

Production statistics, selected for random years, are reproduced in Table 12.7.

In 1989, the country produced 7 900 000 tonnes of bauxite, of which 3 630 000 were consumed locally. It also produced 1 800 000 tonnes of alumina, which gave 888 000 tonnes of aluminium.[92]

Exports of aluminium began in 1978, and massively in 1980; by 1991, some 75% of total production was exported.

The very positive growth of the industry over the last 20 years (a fivefold increase between 1971 and 1981 and a fourfold increase between 1981 and

1991) seems to have reached an end, with production now at a plateau of about 1.1 million tonnes of aluminium per year, no further expansion plans on the horizon, and exports running at a good percentage of total production.[93]

But the industry is subject to the whims of unpredictable governments. With much of the cost of aluminium production being electricity, and all electricity in Brazil coming from government owned hydroelectric plants, prices are a political football, and the wellbeing of the producers depends on the whims of bureaucrats. The industry is likely to pass through a grim decade in the 1990s.[94]

Placer gold once more

It must have been in about 1960 that reports of gold placers on the Tapajós river came out of the Amazon basin, and a gold rush was on. It has yet to end.[95]

In about 1970, gold was found along the Xingú river. Then, in around 1980, Serra Pelada, in northern Pará, was discovered. And then, later on in the 1980s, gold was first found in the Madeira river, then all across Roraima, then again along the Madeira and finally, in 1990, along the upper Negro river, towards Colombia and, in fact, along the border and well into Colombia. And the finds along the Madeira and Mamoré rivers extended, of course, also into Bolivia, just as finds in Roraima extended well into Venezuela and Guyana.

Once again it was all placer gold, very fine dust taken out of the rivers by washing the gravel. But by now very little was obtained by individual lone prospectors, standing hip deep in the rivers and skilfully rotating the conical pan. Modern technology had arrived.

Small dredges, or simply floats with suction hoses, were now the standard equipment. All were provided with gasoline or diesel driven pumps. The simple floats had 10 or 12 inch flexible hoses which extended to the river bottom where a diver – working often as deep as 20 metres, occasionally even deeper – guided the mouth of the hose over and into the bottom sand or gravel, preferably into pockets or holes between rock formations, as there the heavy but fine gold dust was usually concentrated.[96]

The diver's task was hard and dangerous. Depending on the depth, a stay of 30 minutes could be too long and the diver could suffer 'the bends', cramps and nitrogen poisoning. But some of them stayed down as long as two hours. Air reached them through a thin, quarter inch nylon hose, sent down through a small diesel pump that often mixed air with fuel fumes.

Often the water at the river bottom was icy cold, which added to a common feeling of numbness. And there, at the bottom of the river, everything was done by touch, as the Amazon rivers are usually muddy.

The dangers the divers faced were – besides nitrogen poisoning – branches

Panning for gold (by Percy Lau). Source: reproduced by permission of Revista Brasileira de Geografia and of the Fundacao Instituto Brasileiro de Geografia e Estatistica – IBGE.

and even tree trunks drifting with the current at the river bottom, boulders rolling with the current, cave-ins when too much gravel had been sucked up from a particularly deep hole and the walls of the hole suddenly collapsed. Drowsiness was a factor, as was accidental rupture of the air hose. With the distances and the uneven river bottom and the often rather fast current, life often depended on everything just working perfectly – and that was rarely the case.

But by 1991, most of the floats had become dredges, with rigid tubes replacing the flexible hoses, and screws at the tips. The tubes were lowered and guided by derricks, maybe occasionally with an exploratory assist from a diver.

The screw – and the tip of the tube – were now guided gently, yet still by simple guess, along the river bottom with the pump sucking up water and sand and gravel – and maybe gold.

At the surface, on board the dredge or float, inclined tables received the gravel, allowed most of the water to filter out right away and let the larger gravel and stones roll off and fall back into the river. The finer gravel and sand would be gently washed down a 10 or 12 foot channel lined with thick nylon carpeting, often the most expensive and thickest to be had at the nearest town.

It was in this thick – often as thick as 1 inch – nylon carpet that the gold got stuck. It sank to the bottom while mere sand was just washed off.

Every so often the operation was stopped and the carpet cleaned, the fine black muck at the bottom scraped out and kept separately – for 'refining'. The muck often consisted already of as much as 30% gold, the rest usually being illmenite or some other iron ore compound.

The process was by the year 1991 still quite primitive. While exact calculations as to the percentage of recovery do not exist, the number most often bandied about was 50%. It meant that probably not more than half of the gold contained in the gravel ended up as refinable concentrate. But efficiency was not really the objective of the 'miners'. They simply wanted gold – as much as possible and as quickly as possible.

He who had a bit of capital, or simply a bit of credit, or had a reputation for 'getting things done in the wilderness' and was thus worthy of trust, could build a float and move into the river, with three or four companions, some food, some fuel, and the will to work.

Some 9000 such floats and dredges were said to have been operating on the Madeira river – and a few of its tributaries – near Porto Velho alone by the early part of 1991.[97] Practically all were engineless hulks, usually pontoons, hastily and crudely welded together at the river's edge where access roads allowed rigs to take the oversized cigar-shaped pontoons to be taken and discharged. Platforms – decks – would next be built on top of the pontoons, plus a small hut, a derrick or two, an anchor and a few essential utensils. Sleeping quarters for the crew – the operators – would consist of hammocks, and a small dinghy would be essential for contact with the shore.

Shallow stretches of river would obviously be better suited than deeper parts, and a slow moving part was better than a fast section.

As many of the rivers were, and no doubt still are, full of gold, it mattered little where they worked, or how much they understood of dredging and washing gravel. No better teacher than experience when there was gold to be found. And where and when nothing was found, others also found little or nothing, and all soon moved on.

The process of 'refining' the gold-containing mud was – for the early 1990s – also not very sophisticated. It was, in fact, most primitive, extremely wasteful and environmentally destructive.

The most knowledgeable member of a given team would 'estimate' the gold content of the muck – usually the product of repeated scrapings of the nylon carpeting over several days – collected in an old oil drum. Of course, if they cleaned the carpeting too often, the muck was of poor quality, usually much more iron than gold. If they delayed cleaning for too long, and the carpet was already 'full' with gold and iron, additional gold that could not adhere any more to the nylon would simply be washed off.[98]

The 'estimate' of the gold content was intuitive, usually based on panning a handful of muck two or three times to decide, by means of 'occular judgement', that the accumulated muck in the drum would contain, say, 10% gold. The theory was that 10 kilos of muck would contain about 1 kilo of gold, and to extract, or bind, the gold out of the muck would require 1 kilo of mercury. The mercury would then be mixed into the muck, and the mixture stirred. Much better than stirring would be to knead it with human hands, which was usually done, and for hours on end.[99]

In time, distinctive 'balls' of amalgam would form – gold bound by mercury – and would separate from the progressively less gold-containing iron dust. Eventually, these balls would be separated, and the remainder, presumably no more than iron particles, would be dumped back into the river.

As nothing in the entire process was based on analysis, and everything depended on guess work, the final quantities of gold recovered were also simply that, guess work. If too much mercury was used, the excess that did not bind gold was lost, infinitesimally small particles thrown into the river together with the iron. If, on the other hand, too little mercury was used, not all the gold in the muck could be amalgamated and the unbound gold was thrown away with the iron particles.

Worse still was the final step of the refining process, the separation of the gold from the mercury. For this, the usual process was to spread the amalgam, the gold bound with the mercury in a proportion of roughly 50/50, on to a crude 'frying pan', or pizza pan, to a thickness of about 1 centimetre, and then to use a blow torch so that the hot flame would 'burn off' the mercury. It would, in fact, bring about an evaporation of the highly volatile metal. It would dissolve and rise as vapour – looking indeed like white water vapour – to a height of about 2 or 3 metres, before condensing again and settling back down. In this process of first rising as vapour and then settling as droplets, the mercury would settle on any object at hand – including on men's eyebrows and the hair on their heads, their moustaches, even on their forearms – as a form of eery-looking white mist. But most of the condensing vapour would simply fall into the river, or on to the float or dredge, from where it would be washed into the river. Almost nothing of the mercury would be recovered for reuse.[100]

Did the men realize that this kind of careless handling of mercury was dangerous to their health? And that the quantities settling into the river would not degrade, but would stay at the river bottom, perhaps forever, a good part finding its way into fish, and on into humans who would eat the fish?

And the authorities – were they aware of the dangers, the misuse, the abuse and the pollution? And did they do anything about it?

The answers to all these questions were, and remain, ambiguous. The miners – really *garimpeiros* – were generally aware of the dangers of indiscriminate, careless and prolonged handling of mercury. But the threat of illnesses some time in the future caused little fear, when weighed against the possibility of wealth, of a strike, of much gold – maybe tomorrow. And even though tomorrow always remained a dream, it still always seemed just another dawn away.

The authorities were aware of the dangers, but besides occasional pious outcries by a high government official, and one or another well-intentioned law being passed by congress, or decrees being issued by the president, that mercury should not be used and that efficient methods be used to extract the gold, so that the rivers should not be polluted, nothing was done.

Gold and tin *garimpos* are, in fact, in the Brazil of the early 1990s, first and foremost a social problem. If, indeed, some 800 000[101] *garimpeiros* work in the rivers and ravines of the Amazon basin, then probably as many as 4 million direct dependents benefit from this work. If one adds to these the many others who sell supplies, build equipment and transport supplies, probably another 2 million, including their dependents, are tied to this type of mining work. A grand total of 6 or perhaps even 8 million people who get their livelihood from this mining is a very large group within the total population of about 156 million Brazilians.

By 1991, placer gold in riverbeds had been identified all over Brazil and was being pursued in practically all states of the Brazilian federation. The major concentrations were in Pará, along both the Xingú and Tapajós rivers, and then in Amapá, Roraima, Rondônia, Goiás and Bahia. But there were also deposits being worked in Rio Grande do Sul, São Paulo and Rio de Janeiro. Between gold, tin and gemstones, some 1800 deposits had been identified by DNPM, the National Department for Mineral Production. And many were producing early in 1991.[102] At the gold placers something like 100 tonnes were produced.[103]

More or less 100 tonnes is what was registered statistically for 1989 and 1990. By the middle of 1991, production was down. But though gold mining

Table 12.8. Gold production through the centuries

Period	Production (tonnes)
17th century	19
18th century	840
1801–33	86
1834–1900	140
1901–73	335
1974–9	33
1980–90	1044
Grand total	2497

itself is practically tax-free in Brazil, the resulting 'income' is not. And ownership of gold, and wealth, is dangerous in the Brazilian interior because of the risk of being robbed. Thus, many of the small quantities that are produced by one- or two-dredge operators, do not get registered. The men get paid in gold, and they exchange it on for supplies, whores, and larger, more tangible, assets.

There were, by 1989, around 100 gold buyers and melting shops in Porto Velho, the capital of Rôndonia, a city then with 250 000 inhabitants. Only a fool would assume all of them had regular bookkeeping, to register all purchases, sales, smelting and refining charges, and that taxes on all transactions had been paid.

If 100 tonnes was Brazil's registered gold production in 1989, real production, including, in other words, those quantities not reported but smuggled throughout Brazil and sold out of the country, was probably at least 150 tonnes, if not 200 tonnes.[104]

By the middle of 1991, Brazil's federal government was once more looking most energetically into the comings and goings in Rondônia – in gold production and cocaine smuggling. All dredging in the Madeira river was forbidden, all dredges were to be confiscated.[105] It was even said that the gold dredging operations were not really that but just subterfuges for the laundering of cocaine funds.[106] We should read about these investigations and prohibitions with considerable scepticism.

Summarizing gold production in Brazil, reasonably reliable production statistics give the picture shown in Table 12.8.[107] The statistics are those of DNPM, except those for the period 1980–8 which come from *Gold 1991*, a Gold Fields Mineral Services Ltd yearly brochure. But the statistics are only 'reasonably' reliable. They were no doubt diligently and honestly put together, but the factors of smuggling and non-registration remain unaddressed and troublesome. Just as large quantities of gold are not registered today – to avoid production taxes at the source and income taxes for the owners – they were also not fully registered in the past. We should assume that no more than half of any gold production was and is being picked up statistically.[108]

A bright future

Nobody should doubt that Brazil's mining sector faces a bright future – and that means that the country faces a bright future. In terms of the reserves already known and the variety of metals known to exist in the ground, Brazil has probably no rival anywhere in the world.

The reserves of iron and bauxite seem to be limitless, with substantial reserves of manganese ore. Kaolin is available in large quantities, and gold and cassiterite also appear to be. And the areas as yet unexplored or only partially explored are still vast; much will no doubt still be discovered.

Brazil has been, as we have seen, slow or late with its development, always 50–100 years behind many of the other Latin American republics. It should thus not surprise us that the cycle of xenophobia, restrictions on foreign investment in the mining sector and so on, were late in coming to Brazil. The 1988 constitution imposed new restrictions and foreign participation in mining is now limited to 49%. But the regulations and amendments to these constitutional restrictions have not yet been published, and in 1993 the constitution is to be reviewed and revamped once more. The general tendency of the Brazil of the early 1990s is to open the country up, to become more friendly towards foreign capital, and to incorporate itself into the world economy. But there is a big chasm between the outwardly voiced international and *laisser-faire* orientation of the government, and the much more provincial and often even xenophobic orientation of the Brazilian congress.

The 1990s will be crucial for the fundamental orientation of Brazilian society – to join the world, open up and develop, or to turn once more inward and resign itself to second or third rate status among the great nations of the world.

Notes

1. A Frenchman, Jean Cousin, had probably landed on the Brazilian coast already in 1488, and Vincente Yañez Pinzón too, early in the year 1500, before Cabral reached Brazil.
2. Simonsen, page 96. Roberto C. Simonsen's *Historia Economica do Brasil* is another of those basic works any student of mining in the Americas must review. It is an excellent study – unfortunately only covering the period until 1820.
3. Ibid, page 97.
4. Ibid, page 115.
5. Poppino, *Brazil, the Land and the People*, page 123.

6. The extreme to which national pride can go was shown in a newspaper column published on 26 July 1991, *by Folha de São Paulo*, under the signature of José Sarney, former president of Brazil (1985–90). Mr Sarney wrote that the Ianomami are perhaps the oldest people on earth (not just in South America), some 50 000 years old. And: 'If they exist, it is because Brazil preserved them until now, while other people decimated their Indian populations.'
7. Simonsen, page 134.
8. Ibid, page 219.
9. Latif, *As Minas Gerais*, page 35.
10. Simonsen, page 273.
11. Ibid.
12. Eschwege, *Pluto Brasiliensis*, volume I, pages 197 and 203.
13. Ibid, page 53, but principally page 210.
14. Ibid, page 53.
15. Ibid, pages 54–5.
16. All the figures in Eschwege's *Pluto Brasiliensis* are expressed in *oitavas*. One *oitava* weighed 3.586 grams.
17. Eschwege, volume I, page 55.
18. Ibid.
19. Ibid, page 56.
20. It all reads like a repetition of Potosí.
21. Eschwege, page 56.
22. Ibid, page 57.
23. Burns, *A History of Brazil*, page 61.
24. Ibid.
25. Eschwege, volume II, page 87.
26. Simonsen, page 289.
27. Ibid.
28. Eschwege, volume II, page 87.
29. Ibid.
30. One almost suspects that the authorities of twentieth century Brazil have studied the old reports and have decided to repeat past procedures. Thus recently, in 1990–1, various government actions 'expelled' gold *garimpeiros* from their working grounds in Ianomami (native) lands in the state of Roraima.
31. This also has parallels in the Brazil of 1991, also with regard to *garimpeiros* in Ianomami lands.
32. Eschwege, volume II, page 88.
33. João Fernandez de Oliveira was, of course, the one who built the lake. It stands to reason that he must have made money hand over fist, and very much, we may assume, at the expense of the crown. Or maybe it was simply because he had such a sweet deal. And that he held on to the concession until 1771 – another 17 years – 'with partners', speaks of collusion. The auctions of the concessions were probably rigged.
34. Poppino, page 132.
35. Green, *The World of Diamonds*, page 132.

36. DNPM/DEM: 70% were gem quality, the balance industrial diamonds.
37. Sauer, *Brazil, Paradise of Gemstones*, page 11.
38. The author visited the Chapada Diamantina late in 1991 and spent several days exploring the area and visiting several of the diamond dredges still operating.
39. Eschwege, volume I, page 19.
40. See the jacket of *Pluto Brasiliensis* by Eschwege, volume I.
41. 'Os 150 anos de Mina de Morro Velho', *Brasil-Mineral*, June 1984, page 11.
42. Dencker, page 96.
43. Simonsen, page 375.
44. Simonsen, pages 385–6. The figures appear actually as:

	6.090:264$284	
less	239:019$285	(for food imports from Madeira and the Azores)
leaves	5.815:244$999	

And 5.851 in Portuguese spelling becomes 5851 in English.

It may be a bit daring to declare 'imports into Portugal from Brazil' as equal to 'exports from Brazil', but there were then no exports to countries other than the motherland. There was, in fact, no contact, or almost no contact, with other countries. Foreign vessels were not allowed to put into Brazilian ports.

45. Silva Bruno, *Historia do Brasil*, volume 4, pages 80ff.
46. Ibid.
47. Burns, page 141.
48. Poppino, page 260.
49. It was actually Padre José de Anchieta who in a letter dated 1554 mentioned the presence of iron ore for the first time. Magalhães Gomes, in his *Historia da Siderurgica no Brasil*, page 40.
50. Eschwege, volume II, page 201.
51. Ibid, page 202.
52. Eschwege's text, as translated into Portuguese, which is the version I used, shows some humour. He calls Manso a 'multatto by birth', as if another kind of mulatto could exist, and then goes on, 'who, having read in a book about iron-making, followed the very Portuguese and very Brazilian habit of then thinking this would make him an expert in iron-making.'
53. Eschwege, volume II, page 203.
54. Ibid, pages 204–5.
55. Magalhães Gomes, page 50.
56. Ibid, pages 50–1.
57. Ibid, page 75.
58. Ibid, page 61.
59. Ibid, page 67.
60. Ibid, page 50.
61. Eschwege, volume II, page 245. This is not part of Eschwege's text (Eschwege, after all, wrote his book in the early 1830s) but comes from a 'Note of the translator' who, skilfully, completes the information for the reader.
62. Ibid.

63. Magalhães Gomes, page 139.
64. Eschwege, volume II, page 245.
65. Magalhães Gomes dedicates an entire chapter, Chapter IV, to the subject.
66. Ibid, page 141.
67. Ibid, page 146.
68. Ibid, page 189.
69. Ibid, page 201.
70. Ibid, page 271.
71. *Folha de São Paulo*, of 30 July 1991, quoting Ilafa, Instituto Latino-Americano de Fierro y el Acero.
72. Where this is done depends on the availability of such forest, the degree of control or lack of control by the authorities over such activities (essentially forbidden) as well as road conditions and distances to consumption centres.
73. These statistics come from Abracave, Anuario Estadistico 1990 and courtesy of Christoph Behn, of Metallgesellschaft do Brasil, São Paulo.
74. Poppino, page 251.
75. Booz-Allen & Hamilton do Brazil, *Minerals & Metals Outlook Brazil 1977.*
76. Wang and Yih, *Tungsten, Metallurgy, Properties and Applications*, pages 46–7.
77. The writer travelled to these wolframite *garimpos* twice, in 1981. Several hundred *garimpeiros* must have been working there then, and subsequently the numbers must have increased, for a while, before lower tungsten prices terminated the activities.
78. My trips were in every respect simply exhilarating and, but this mostly in retrospect, scary. Flying in a single engine plane, landing on a small strip, walking for hours through the jungle, and then staying a few days in the *garimpo* was plain adventure, as was the return flight into the night, with jungle burning below and smoke clouds reducing visibility to almost zero.
79. The story of the origins of tin mining in Rondônia comes from Eric de Champeaux de la Boulaye, a French count who settled in Porto Velho in the 1960s as the ore buyer of Cesbra. He has probably forgotten more old lore than others remember. He was also the one who related the tale about Dr Cassio de Souza Mello, who would himself never have mentioned the story.

    I met Dr Cassio for the first time in 1976, in Rio de Janeiro, and thereafter it was he who during some 20 joint trips to Rondônia introduced me to cassiterite mining. It was then that I became an unashamed admirer of Dr Cassio, of his knowledge of tin mining, and his understanding and love of the Brazilian mining industry in general and the Amazon lowlands in particular.
80. My own association with the industry lasted from 1976 to 1980. The Santa Barbara mine then belonged to Mibrasa, and Mibrasa was owned 50/50 by Patiño and Philipp Brothers. Both eventually sold their interests and withdrew from the industry.
81. Once more, the Metallgesellschaft statistics.
82. Kinch, 'Bugbear of the world tin market, *Metal Bulletin*, 20 August 1990.
83. Ibid.
84. But some of the ups and downs in the mining industry are amazing, and there

can suddenly be a dramatic increase followed by an equally dramatic decline. By 1991, Brazil's tin production was very suddenly down again, to about 30 000 tonnes, and exports to only some 22 500 tonnes. The established producers – both miners and metal-makers – and many politicians had finally managed to prevail upon the government to close the Bom Futuro *garimpo* in Rondônia, and as that had for a while accounted for as much as 10 000/12 000 tonnes a year, production was drastically down. Lower prices accounted for the rest. But the ore is still there and higher tonnages will probably soon be seen again.

85. Santos, 'Carajás – historia e perspectivas'. I was given a copy of this report by Diana Kinch, the Latin American editor of *Metal Bulletin*. The story was independently confirmed by Dr Celso Pinto Ferraz who, in 1991, was professor at UNICAMP, the São Paulo State University in Campinas. A number of years ago, Dr Pinto Ferraz was working with DNPM, in Brasilia, and he was present at one meeting between the board of directors of US Steel and the board of directors of CVRD, including the (then) mining minister Shigeaki Ueki, when CVRD proposed that the partners (US Steel and CVRD) invest something like $100 million dollars each, which sum, however, US Steel was not willing to put into iron ore mining at a time when prices were low and the prospects for better business bad. In the following exchange the idea of Brazil buying out US Steel was aired and when US Steel appeared willing to sell, a letter of intent was at once prepared and signed.

    Dr Pinto Ferraz also pointed out to me that there was nothing unusual in US Steel having formed a partnership with CVRD in 1970. It was, he says, quite the custom in those years, the 1960s and 1970s, for foreign companies to seek an association with the government, or a government company, in the host country. It would, he felt, have been quite difficult for US Steel to develop successfully the biggest iron ore mine in the world, in those years, in a Third World country, without major political objections. And, indeed, it had only been two years earlier, in 1968, that the military in Peru expropriated the US owned International Petroleum Company.
86. Kinch, 'Iron ore stands at production', *Metal Bulletin*, 20 August 1990.
87. The sources for the Genesio Ferreira story were innumerable newspaper articles.
88. Many excellent photos of the pit were published over the years; I got my information and the descriptions from several *garimpeiros* who had spent time at Serra Pelada.
89. My own guess.
90. This is a guess, probably a rather intelligent one. It has been made again and again by many people. It is a generally accepted rule of thumb.
91. This is my guess, but it is based on comments received from others in the industry.
92. CIEF, Centro de Informacões Economico-Fiscais, and Cacex.
93. Kinch, 'Energy costs rein in primary aluminium growth', *Metal Bulletin*, 20 August 1990.
94. My own judgement. Diana Kinch, the Latin America editor of *Metal Bulletin* wrote to tell me that Abal, the Associação Brasileira de Aluminio, claim that

the Brazilian aluminium industry's energy costs are the highest in the world, currently at 27 mills (27 × 1000th of a dollar per kilowatt hour) and that this compares to costs of 8.8 mills in Venezuela and between 5 and 7 mills in Canada; also Abal apparently claims that at current low world aluminium prices of around $1200 per tonne, Brazilian aluminium producers are all making losses since their energy breakeven level is at a maximum of 20 mills (letter dated 2 October 1991).

95. This is basically my own conclusion. The reports that were published in 1991, and the opinions that were voiced even by those in the industry, were pessimistic, and there was even talk that the *garimpeiro* business and the gold rush were over. Maybe it was, temporarily, in 1991.

    But gold always was subject to its price, and the gold price had not moved for a year or two (looking at it from the vantage point of early 1993). Furthermore, the gold boom in Brazil of the 1980s had partly also been fuelled by a very confusing pricing policy, and by a weird exchange rate policy.

    The Brazilian gold price obviously always depended on the international gold price, expressed in dollars and then converted into local currency. For a while, in the 1980s, the parallel (or black) exchange rate for the dollar (and the rates for other freely convertible and strong currencies) was running at about 70–100% above the so-called official exchange rate. And as gold was traded, unofficially and officially, at a cruzeiro price converted at the parallel exchange rate, he who sold gold received, obviously, 70–100% more cruzeiros. Or, in other terms, the cost of his materials, supplies, and often also labour, was reduced by almost 50%. And then when the Collor government, beginning in about March 1990, pushed the parallel exchange rate down to about 10% above the official exchange rate (now called the free rate), the income of the gold producers, expressed in cruzeiros, went down sharply. And thus, for a while, the boom was down because the bloom was off. But, basically, in my judgement, the Brazilian gold rush is still very much alive.
96. Observations during my own trips.
97. Many newspaper stories, probably exaggerated.
98. My own observations.
99. My own observations.
100. My own experiences, during several trips in 1987 and 1989, as a result of which I even obtained an official government licence to pan for gold in Rondônia, a document I still own and cherish.
101. Kinch, 'Gold garimpo era over – DNPM', *Metal Bulletin*, 30 May 1991.
102. Most of the 800 000 *garimpeiros* were busy working the high grade claims.
103. Double or triple what the official figures show.
104. My own estimates.
105. Many newspaper articles, July 1991.
106. The Brazilian newspapers were full of these stories in July and August 1991.
107. *Brasil-Mineral*, June 1984, page 16, citing DNPM, SICEG.
108. My own guess.

CHAPTER 13

Canada

Mining and metal-making are fundamental to Canada's economy. Also this country is one of the great mining areas of the Americas and of the world. Its mining industry is of truly gigantic proportions.

Slow beginnings

The discovery and settlement of Canada began quite early. But the early European travellers did not find metals and ores.

John Cabot, an Italian with a British ship, found ocean areas rich in fish near some land he called 'new found land' in 1497. Much fishing took place there subsequently, by British, French, Dutch, Spanish and Portuguese fishermen.

In 1535, a Frenchman, Jacques Cartier, sailed up the St Lawrence river, all the way to where Montreal is today. It was then that the fur trade began. And that is what the early visitors found and exploited: fish, wild animals with furs and, subsequently, timber. But no minerals.

One M Prévert, a Frenchman who accompanied Samuel de Champlain, is credited with the first mineral discovery in Canada. He found copper ore in about 1604, in the Bay of Chaleur, in Quebec's Gaspé region.[1] But it was to be several centuries before the place where the sample came from became Gaspé Copper Mines Ltd and copper mining became a major industry. Also in 1604, Master Simon, a mining engineer in Champlain's group, discovered iron and silver ore at St Mary's Bay in Nova Scotia.

In 1654, King Louis XIV granted a concession to one Nicholas Denys, to mine gold, silver, copper and other minerals on Cape Breton Island, Nova Scotia. But it was only in 1672 that Denys found coal there (the implication being that the original concession had just been a general licence, not a permit for a specific mineral deposit); it then took until 1720 before coal was produced. Finally, four years later, in 1724, coal was exported to Boston.[2]

After very slow beginnings in the seventeenth and eighteenth centuries, Canada eventually became a crown jewel of the British Commonwealth and then a wealthy nation in its own right – in terms of mining.

It was first known for its furs and timber, and then for its grains. Today, Canadian nickel is essential for world supplies, and Canada's gold, silver, platinum, lead, zinc and copper play vital roles in the supply patterns of the world.

Contrary to Mexico, Peru, Bolivia and Brazil, where gold and silver were the first metals to be discovered and worked, in Canada it was iron ore – and iron-making – that got the industry started in about 1733.

This should not lead us to believe that gold and silver deposits did not exist in Canada, or that they were not worked, but little information on this has come to us.

Those who penetrated into the interior may have looked for gold, but they were not really mining men, or prospectors. And the concentration on the fur trade probably precluded a more earnest and more systematic search for metals. It was thus iron that was worked first.

In 1733, iron ore was smelted for the first time in Canada, in Trois-Rivières, Quebec, at the Forges St Maurice. The plant worked off and on until about 1830.[3]

By the year 1800, a first blast furnace was erected at Lyndhurst, Ontario, at a waterfall on the Gananoque river. But this prospect did not prosper, because of the poor quality of the ore. It was abandoned after a few years.

One Joseph van Norman was more successful, first with a blast furnace in Charlotteville Township, Ontario, in 1820, using a small furnace first erected in 1813 but which had not produced more than 2 or 3 tonnes in the intervening years. Van Norman made some modifications and produced 700–800 tonnes a year until 1847 when ore and fuel supplies were exhausted.

Van Norman next moved to a town called Marmora, set up a new furnace and began once more to produce pig iron – only to fail again.

When new ore deposits were discovered near Charlotteville Township, he moved back there in 1854 and built a new blast furnace. But he failed once more as the iron could not be used for railroad wheels.

The story of Joseph van Norman warranted telling as his attempts – successes and failures – to produce commercial grade iron lasted some 35 years, and show how much trial and error – as well as persistence – were required for the ore and metals industry to progress.[4]

By 1851, the Geological Survey of Canada, which had been instituted back in 1845,[5] had identified numerous, and among them several quite large, iron ore deposits. By 1857, exports were begun via the Rideau canal and across Lake Ontario to the United States – to the steel works in Pittsburgh, Pennsylvania.

By then, and actually as early as about 1823, a few gold placers had been identified,[6] on the Chaudière river in Quebec, at Mitchell Harbour, on Queen Charlotte Island in British Columbia, and at the mouth of the Pend'Oreille river, also in British Columbia.

In 1744, silver–lead ores had been found on the shores of Lake Temiskaming in Quebec, and in 1770, native copper at Point Mamainse, on the north shore of Lake Superior. In 1771, a clerk of the Hudson's Bay Company, Samuel Hearne, had begun prospecting for copper in the Northwest Territories.

By 1847, copper deposits had also been identified in Eastern Townships, Quebec, as well as silver near Thunder Bay on Lake Superior.

Gold, and the Klondike

The first significant finds of gold were in the Cariboo and Yukon regions. It was the crew of a Hudson's Bay Company ship which found gold bearing quartz veins on Queen Charlotte Island – running 26% gold for some 80 feet on the surface before disappearing under overburden. In 1858, another Hudson's Bay Company ship brought some 800 ounces of gold to San Francisco from the Thompson and Fraser river country in British Columbia. A rush was soon on.

The first placer gold in the Cariboo was found in 1859 by one Peter C Dunlevy – in a bend of the Horsefly river. The richest creek in the Cariboo was discovered in 1861 by 'Dutch Bill' Dietz; it was named Williams creek after Dietz. It supposedly yielded more gold than any other stream in British Columbia, or New Caledonia, as it was then called. Dietz himself staked out one of the poorer claims and eventually died in poverty.

By 1863, gold production in British Columbia had already begun to drop off again and the prospectors moved on, north to the Yukon. That there was gold in the Yukon gravels was known, but it took a while before it was considered an economic proposition – and really before a few spectacularly rich finds caused another rush.

In 1872 an Irishman, Arthur Harper, went with a group up the Yukon in pursuit of native copper. Good gold prospects were found at the mouth of a tributary which they called Forty Mile. They passed it up, discouraged by talk of a dangerous canyon. Forty Mile creek later, from 1887, became one of the big gold producers.

The Klondike discovery itself was made by one Robert Henderson. He found Bottom creek. George Washington Carmac and two Indian friends, Skookum Jim and Tagish Charlie, found and staked Bonanza creek – all in

1896. The Klondike rush was on. It was to last some seven or eight years. Production rose from about $300 000 to about $23 million in 1900, the peak year. By 1903, production was already down to about $12 million. The cumulative total value had by then reached about $100 million.

But it was not really mining that went on during these early rushes.

The first discoverers of gold in the vast interior of the north-western part of the continent may well have been *promyschlenniki* – the Russian trappers who went east after the rich pickings of Siberia appeared exhausted and Czar Peter the Great (1672–1725) ordered a search for new sources. The Russians went to Alyaska, as the natives called their land.[7] Or it may have been French or British trappers pushing west from Ontario who discovered the gold.

In any case, the first gold in or on or near the Yukon river was apparently discovered in around 1869.[8] It then took until 1896 before a real strike was made.

By then, Russia had sold Alyaska, or Alaska, to the United States in 1867 for $7.2 million. The high price was criticized by much of the American public.

Why am I discussing Alaska and the United States? Are not the Klondike, the Yukon river, Dawson, and the smaller creeks nearby, where much or most of the Klondike gold was found, part of Canada, and not part of the United States? Indeed! Yet it is not that simple.

Basic geography explains quite a lot. The Klondike river runs into the Yukon river and most of the Yukon river runs through Alaska. It can thus be said that the Yukon is basically an American river. And Dawson is less than 80 kilometres from the border of Alaska, and about 1800 kilometres distant from Vancouver, BC and Edmonton, Alberta. Then – and now. The distances were close to unsurmountable then. In fact, the closest contact point with civilization was – then, in about 1895 – the port of Skagway, on the Lynn inlet (then, and now, part of American Alaska), and the early prospectors walked from that small, and essentially primitive, port up the White Pass (or walked from nearby Dyea – another, even smaller port – up the Chilkoot Pass) to the high plateau.[9] They then went on to Lake Bennett, from where they boated, if they had a boat, or floated, if they had a float, down the Yukon, across or around several rapids. That was the shortest and quickest route to Dawson City and the Klondike.

Skagway and Dyea were still on American ground; Canadian territory began only at the top of the passes. And at the beginning there were no American authorities there. The red-coated 'North-West Mounted Police' could only control the route from the passes onward.

Skagway and Dyea became synonymous with the old Wild West, with lawlessness, wanton murder, corruption and gangs. One Jefferson Randolph Smith, 'Soapy Smith', a notorious cheat and gangster from Seattle became, for a while, the uncrowned ruler of Skagway.[10]

The road, or rather path, up the passes from Skagway and Dyea was about 80 kilometres long to Lake Bennett. Parts of the way, the beginning ascent, were about 45 degrees steep. Some of the gold rushers carried as much as 100 lb up this incline and made as many as 30 trips, to get all their supplies and gear to the top. Conveyances there were none, just the men's backs, and mules and horses.[11]

The animals suffered even more than the men as they were usually overloaded. The way was soon called 'Dead Horse Trail'. On a few occasions, dynamite had to be used to clear the way of the many rotting cadavers.

The men were somewhat better off simply because when a load became too heavy, they chucked more and more until the weight became bearable.

From Lake Bennett it was still some 1000 kilometres to Dawson.

The saga of the first few years of the gold rush of the Klondike is one of triumph and tragedy – but more tragedy than triumph. Once more, as elsewhere, many tried but few succeeded; few became wealthy. Many worked hard, agonized, stayed poor and suffered.

Dawson, and the Klondike, have only a few months of summer a year, and many months of bitterly cold winter. The rivers freeze. Many of the winters last from October to April. Dawson City was then no more than a handful of tents and huts.

Supplies, mostly food, reached Dawson – as well as Fort Yukon downriver and Fort Selkirk upriver – by flat-bottomed boats. These could only operate when the rivers were not iced up. No supplies meant hunger, as did too heavy an influx of gold rushers, which is what happened in 1897 and 1898.

And he who in 1897 did not get to Dawson in time before the river froze up was stuck at Lake Bennett until the end of May of 1898.

J J Pitts, the store keeper at Fort Selkirk, still 300 kilometres upriver from Dawson, where the Yukon river actually begins (formed by the coming together of the Lewis and Pelly rivers), kept track of the prospectors who passed through the fort: 7124 until June of 1898, with 15 000 tonnes of supplies.

And still they came, from Skagway and Dyea, up the passes and down the Yukon. They also came via the 'All-American Way', from Port Valdez, across the glaciers, then up the Copper river, down the Tanana river, and finally up the Yukon river.

The 'back door' way was by boat up the Alaska coast and then up the Yukon, comfortably, all the way to Dawson. Comfortable, that is, if and when the vessel did not run on to a shallow sand bank, or get caught in ice, in which case it had to be abandoned.

Another 'back door' route was supposedly the overland way from Edmonton. Many wanted to take it – and the population of Edmonton rose from 700 in the autumn of 1897 to over 10 000 by July of the following year. The trek to Dawson City was supposed to take only 90 days. But there was no route, no path, no trail. It was simply plain wilderness, or rough wilderness, crossing the Rocky Mountains, going down rivers, across or around unnavigable rapids. Some 4000 men made it, out of a total of maybe 20 000. The trek took two years.[12]

Individuals from all walks of life rushed into the wilderness to strike it rich. Sluice box and rocker were the basic equipment – and a pick and a shovel – at the beginning, to work the shallow deposits. Most of what they found was coarse gold; finer material was worked with mercury into amalgam which was then roasted to drive off the mercury.[13]

By July 1898, some 10 000 claims had been registered. In 1899, Dawson burned down. And by 1903, the Klondike gold rush was basically over.

And though Dawson City, the Klondike river and the many gold bearing creeks are all in Canada, and the Klondike gold rush is essentially a Canadian mining story, it was in reality mainly an American mining story as most of the miners were US citizens.

COMINCO

The history of COMINCO – the Consolidated Mining & Smelting Company of Canada Ltd – began in 1906. As a subsidiary of the Canadian Pacific Railroad, it has become an important industrial component of modern Canada.

In the 1880s, substantial base metal mineralization was found in the Kootenay district of southern British Columbia. What was originally just a prospectors' camp, Rossland, quickly became a mining town. The first claims were staked in 1890, by one Joe Moris in partnership with a Joe Bourgeois. Among their claims were Centre Star and War Eagle. An extension of the Centre Star claim was called the LeWise claim which Bourgeois gave to one E S Topping, the deputy mining recorder at the town of Nelson, BC, in exchange for Topping paying the recording cost for the other claims. Topping took one look at LeWise, renamed it LeRoi, headed for Spokane, Washington, and formed a consortium to set up the LeRoi Mining & Smelting Co of Spokane. By 1891, a 60 foot shaft and numerous open cuts had been dug. Ore assays ran between 5 and 20% copper, 3 and 10 oz of silver and \$50 and \$500 of gold per tonne. But even at these values, production allowed no profit, as shipping involved ten kilometres by mule to the Trail creek landing on the Columbia river, thence by steamer to Little Dailes, and then on to Butte, Montana, for smelting. The obvious answer was to build a new smelter nearby. Topping chose as a site for a town the Trail creek landing, and he called it Trail.[14]

In 1895, an American capitalist called Fritz August Heinze set up the British Columbia Mining & Smelting Company which acquired land at Trail, built a smelter and a tramway from the mines at Rossland, and negotiated a smelting contract with LeRoi Mines. It took less than a year for the first furnace to be blown in, on 1 February 1896.

In 1894, the original stakers of the War Eagle claim, Moris and Bourgeois, disillusioned with poor results, sold the War Eagle for $17 500 to a group headed by one Patsy Clark. The new owners changed the direction of the drift, promptly to hit an enormous ore body. In 1895, War Eagle paid a dividend of $27 000.[15]

Altogether, Joe Moris was not very lucky. Already in 1892 he had sold the Centre Star to one Oliver Durant who, in 1895, formed the Centre Star Mining & Smelting Company.

The LeRoi company next found more ore, in 1896, and decided to build its own smelter which began functioning in October of 1897, at Northport, Washington.

The railways now became important. Canadian Pacific Railway (CPR) wanted to bring its line through the Crownsnest Pass, across the Kootenay country. It would compete with the Heinze interests which were planning a Columbia and Western Railway. There was much back and forth for a while, as was the custom in those days of early and ruthless capitalism, and then Heinze sold out and CPR ended up owning not only the Heinze railroad rights but also the Trail smelter. In 1906, the operation was incorporated as the Consolidated Mining & Smelting Company of Canada Ltd, COMINCO, by merging with the Centre Star and the War Eagle Mining & Development Company. In 1911, COMINCO bought LeRoi.

Among the many other fascinating stories of the Canadian mining industry are the origins of the Granby Consolidated Mining, Smelting & Power Company Ltd of Grand Forks, BC, set up in 1899 by eastern Canadian interests, headed by one S H C Miner, a rubber footwear manufacturer of Granby, Quebec. Another story is about the discovery of the Sullivan mine – still today one of the main supply sources of COMINCO. It was discovered in 1892 by a Walter Burchett who was prospecting together with Pat Sullivan, John Cleaver and Ed C Smith. Sullivan died in a cave-in shortly after the staking. The partners worked the mine for a while without much success. They eventually sold it to the LeRoi group; COMINCO acquired it in 1913.[16]

The Flin Flon region on the Saskatchewan–Manitoba border has supplied much copper, gold and silver sulphide ore to the COMINCO smelter at Trail. The complex ore body was first discovered in 1915 but it took until 1930 before the metallurgical problems of how to treat the complex ore were overcome and exploitation could begin. What was originally Harry Payne Whitney's Complex Ore Recoveries Company became, in 1930, the Hudson Bay Mining & Smelting Company Ltd which, again in time, came to be controlled by the Anglo-American group of South Africa.

Noranda

Noranda is another big name in Canadian mining. In 1911, one Ed Horne went east from Kirkland, Ontario, to the region of Rouyn Township and Lake Osisko, Quebec. He liked the geological configuration there but found no outcroppings that excited him. He returned in 1914 for another look. Samples gave nothing. A further trip in 1918 once more brought negative results. Nevertheless, Horne and a few friends were still sufficiently excited to set up a group – the Tremoy syndicate – to put up money and to go to work. Operations commenced in 1920. Assay results finally confirmed Horne's longstanding faith in the region – values ran up to $75 per tonne of ore. Two mining engineers, Thomson and Chadbourne, now headed a syndicate that negotiated the rights to the Tremoy property for about $320 000, payable over five years. The result was the incorporation of a company called Northern Canada, or Norcanada, which through a typographical error became Noranda. Ed Horne returned to cattle farming in Nova Scotia.

By 1924, Noranda had determined seven distinct ore bodies. The big question now was how to connect Rouyn to the outside world. The Canadian National Railroad was persuaded to connect the mines to its main line 45 miles away. The company then decided to go ahead with the construction of a 500 tpd smelter. In time, more and richer ore bodies were discovered and nearby properties added, to make Noranda one of the mining giants of present-day Canada.[17]

The Sudbury basin

Some 250 kilometres south-west of Rouyn is the Sudbury basin, home of the International Nickel Company, INCO, and of Falconbridge Nickel.

Nickel has been a relative latecomer to the metal scene. In 1887, total world consumption of the metal was only about 1000 tonnes, and there was not much of a market. To separate nickel from copper, with which it is usually associated, was still very difficult. Earlier attempts in 1770 and 1846 to work the Sudbury mines, originally known for their copper, had failed. Highly visible outcroppings of copper ore again and again caught different people's attention, but it was only in 1883, after a rock cut by the CNR, some 5 kilometres north-west of Sudbury, had exposed a heavily mineralized zone, that a group of people bought the rights to 310 acres of ground for $1 per acre. It was to be called the Murray mine. And the director of the local office of the Geological Survey pronounced the deposit one of the most promising in Canada. The mine still operates today, but at a depth in excess of 5000 feet.

Other ore bodies in the same basin were identified, such as the Stobie, Victoria and Levack mines. But real exploitation did not result quickly. In

1886, American interests set up in Akron, Ohio, the Canadian Copper Company which began to produce ore thought to contain 7% copper. The ore was shipped to three smelters, the Orford Copper Co of Constable Hook, New Jersey, the Nichols Chemical Company of Laurel Hill, Long Island, and H H Vivian and Co of Swansea, Wales. All three smelters rejected the ore as unusable because 2.5% of the 7% copper content was actually nickel. The Canadian Copper Company thus had to build its own smelter at Sudbury. It was the burning off of the sulphur which began in 1888 that gave the region around Sudbury the eery, moonscape-like aspect which impresses – or depresses – the casual visitor today.

With the problem of economic roasting and smelting solved with the first furnace operation in Sudbury, two problems remained: how to separate nickel from copper and how to create a market for nickel. The second was solved first, in 1889, when the Steel Company of Scotland Ltd, of Glasgow, announced successful experiments with nickel steel armour plate for battleships, which most of the world's navies were just then beginning to build. And in 1892, the Orford works discovered accidentally how to separate molten nickel from molten copper. Three years later, Canadian nickel dominated the world market.

INCO, the International Nickel Company, came about in 1902 when Robert M Thompson, president of Orford Copper, incorporated it by merging Orford Copper, the refinery, with the Canadian Copper Company and other companies he had acquired, such as American Nickel Works, the Anglo-American Iron Company and the Vermillion Mining Co. He also controlled the Nickel Corporation Ltd of England, and the Société Minière Caledoniense. In other words, Thompson for a while had 'a handle' on world nickel supply.

Shortly after the merger, a new nickel–copper alloy, called monel, was developed. It was saltwater and sulphuric acid resistant, and demand for nickel increased further.

Not all who worked in the Sudbury basin were successful. H H Vivian & Co of Swansea acquired and operated the Murray mine, but its smelting techniques were inefficient and it went bankrupt. More successful was Falconbridge.

In 1926, two executives of a mining company called McIntyre-Porcupine, Thayer Lindsley and Joseph Errington, acquired some land in the Sudbury basin, next to Vermillion Lake. In 1928, they and others formed Adventures Limited, soon changed to Ventures Limited. Ventures next incorporated a new company called Falconbridge Nickel Mines Ltd by joining forces with Sudbury Basin Mines Ltd and Falconbridge Mines Ltd, which owned property in Falconbridge Township, originally, in 1901, discovered by Thomas Alva Edison.

Shafts were sunk, a smelter built, a railroad spur laid, and production began in 1930.[18]

The Ventures group went on to become Consolidated Sudbury Basin Mines Ltd, which in 1960 merged with Giant Yellowknife Gold Mine. The Ventures group also controlled or owned part of Sherritt Gordon, United Keno Hill and many others, for a total, at one point, of 185 companies.[19]

And always more . . .

There are very many companies which played and continue to play a significant role in the mining industry of Canada. Teck Corporation, Dickenson Mines, Dynasty Explorations Ltd, and International Mogul Mines are just a few more names.

Canada is today one of the principal world producers of most base metals, also gold and silver, mercury and iron ore, as well as uranium.

By the early 1990s, Noranda was still around, stronger and larger than ever, and now a 'diversified natural resources company', operating in mining and metallurgy, forest products, energy, and manufacturing. In 1990, sales topped $9.5 billion, more than three times those of 1980, but earnings throughout the 1980s had been, well, spotty, with four years, 1982–1985, showing losses. A subsidiary, Noranda Minerals, Inc, owned 50% of Falconbridge Limited (which in 1990 accounted for 65% of the earnings of Noranda) and 51% of Kerr Addison Mines Ltd.

Cominco's sales in 1990 exceeded $1.4 billion, with the Trail metallurgical operation accounting for half of all revenues, Highland Copper for another 25%, and the Polaris mine for a further 15%. Highland Copper and Polaris were apparently the principal money makers.

Notes

1. Jones and Lonn, *Historical Highlights of Canadian Mining*, page 3.
2. Ibid, page 261.
3. Ibid, page 4.
4. Ibid.
5. Ibid, page 262.
6. Ibid, page 261. The text says 'discovered by a woman'.
7. Jeier, *Der grosse Goldrausch von Alaska*, page 9.
8. Jones and Lonn, page 263.
9. The way up from Skagway was actually up the White Pass. Five kilometres away was the Dyea Bay from where the much steeper but also much shorter way was up the Chilkoot Pass.
10. Jeier, page 51.
11. Ibid, pages 72–5.
12. Ibid, pages 158–9.
13. Sounds familiar? Just like Rondônia in 1989! Also See Jones and Lonn, pages 10–20.

14. Ibid, pages 20–1.
15. Ibid, page 22.
16. Ibid, page 25.
17. Ibid, pages 41–6.
18. Ibid, pages 59–65.
19. Ibid, page 66.

CHAPTER 14

The United States

The United States was another sleeping giant and consequently a relative latecomer to the world mining scene. The emergence and growth of its industry eventually came about as from about 1850, like an explosion, to then quickly make the United States one of the most important mining countries in the world.

The origins were small, slow and sketchy.

It is not known whether the pre-Columbian inhabitants of what are today the continental United States did any mining anywhere. One or the other deposit of colourful-looking rocks may have been known, and worked on a minute scale for colouring agents, but they would not have been many. And not even gold and silver deposits were known to them. Yes, legends reached the Spaniards of 'the Seven Golden Cities of Cibola', 'there, to the north' (of Mexico), somewhere in what are today the states of New Mexico, Nevada, Arizona and Colorado. But though a number of conquistadors set out to search for this other Eldorado, nothing was ever found.

There were, in the original 13 colonies, many small copper and iron deposits which were worked, presumably on a very small scale, at the time of the Revolutionary War. And even further back, about 1664, copper was supposedly mined near Lynn, Massachusetts. And, yes, a few 'Indians' were in time found to carry copper ornaments, made from native copper found, not mined.

Other significant dates are 1687, when the lead mines of Missouri were discovered; 1846, when iron was first produced from the extensive Lake Superior deposits; 1848, when gold was discovered in California; 1859, the year of the discovery of the Comstock Lode and, finally, 1863, when lead was found in Bingham Canyon, Utah, which was later to lead to the great open pit copper mine.[1]

North Carolina and Georgia

But in the beginning, albeit again a late one, gold was once more the metal sought. It was in 1799, the same year in which George Washington died (and about 100 years after the gold rush of Minas Gerais, Brazil, 50 years before the California gold rush, and 100 years before the rush to the Klondike) that a young boy, the son of a German immigrant Conrad Reed, found a piece of yellow rock weighing 8 kilograms on the family farm in Cabarrus County, North Carolina. Conrad's father thought the rock was indeed pretty – and kept it as a doorstop. Which is where it stayed for about three years. Only then did a travelling jeweller notice it, and paid Reed $350 for it. It was almost pure gold.[2]

The news quickly spread, and many now began frantically to look for more doorstops. And indeed they found more – the largest weighed 12 kilograms.

Boom towns now rose, and immigrants from all over the world rushed to North Carolina which, together with Georgia, was then part of the western frontier of the emerging United States.

Eventually some 100 mines were worked in a radius of 30 kilometres around the town of Charlotte. It was a real gold rush.[3]

Indeed, so much of the precious metal was found that a branch of the Federal Mint was established, in 1837, where Quarter Eagle and Half Eagle pieces were minted. It worked until 1861 when North Carolina seceded from the Union and the mint closed. It never again opened as a mint, only as an assay office.[4]

With the advent of the Civil War, gold mining declined, yet it continued and sustained the local economy even during the post-Reconstruction times.

The gold mined in the area probably did give Charlotte and the area around it wealth and development, and attracted artisans and business people. This in turn made the area prosperous well into the twentieth century.

By 1849, the California gold rush was on, and attracted national attention. The Piedmont area – as the region around Charlotte is generally called – which until then had been the primary gold source of the country, slid into second place.

In 1829, gold was discovered in what was to become the town of Dahlonga, Georgia. This discovery also resulted in a minor early rush. But first there was the problem of the natives to overcome.

Much of the land where the gold was discovered belonged to the Cherokee tribe, or nation. They had originally lived in Tennessee (about the year 1600), but after a disastrous smallpox epidemic (about 1650) and the erroneous political decision to side with the British during the Revolutionary War, the Cherokee had been pushed into Georgia where they had, however, established themselves peacefully and successfully. But by 1802, Georgia was trying to push them out, and by 1829 their luck had run out. Skirmishes – both legal and physical – continued for another six years, but in 1835, the last Cherokee trekked west, to resettle beyond the Mississippi river. Four thousand of them died during the trek.

With the Cherokee out, the Georgia gold rush was on. The US government even established another branch mint in Dahlonga.

And Manhattan . . . ?

The Dutch settlement of Niew Amsterdam, on Manhattan Island, had, of course, been set up much earlier, in 1625.

The question is, was gold found there?

The reports are sketchy. It was Peter Minuit, the governor of the Dutch colony who in 1626 gave all of $24 worth of cloth, beads, knives, and food to the natives for title to the island and much of the surroundings.

The Dutch were, of course, looking for gold, silver, copper and precious stones. But nothing was found.

Finally, in 1645, a native was seen wearing what appeared to be metallic paint. The dust from which the paint was made was melted down and turned out to be a metal – gold worth 3 guilders.

A search party was sent out into the wilderness to get more, and the resulting bucketful of ore was shipped to Holland. But the ship sank and the sample was lost. The same fate befell a second sample sent somewhat later. Experts were eventually sent from Holland to investigate on the spot; what they found were pyrites.[5]

Did the Dutch really find some gold on Manhattan Island? They may well have! But if so, it was very little. And never again was any found.

And if there indeed had been a mine, or a hole, or an outcropping, or a placer – where would it have been? On 14th Street, corner of 3rd Avenue? Or was it corner of 2nd Avenue? Or further uptown? Maybe the old riverbed where the flakes or specks were found still exists, somewhere in Central Park? Could we imagine a gold rush to some usually ignored spot in the Park?

California, here we come . . .

The importance of the 1848–52 California gold rush goes much beyond just another story of mining. But, strictly speaking, the actual rush was not even mining. During the first two or three years, it was simply a story of a few lucky people picking gold literally off the ground.

The name of a Swiss man, Johann Augustus Sutter, stands at the beginning.

He was a bankrupt merchant who had fled from his creditors, wife and four children to try his luck again in a new world.[6] He had built a little fort in the Sacramento valley to try his hand at growing crops. It was one of Sutter's foremen, a carpenter from New Jersey by the name of John Marshall who, sent up the American river to the Coloma valley to build a sawmill, discovered gold – in nuggets and flakes – at the river bottom. It was 24 January 1848.

Before the news could spread, Sutter and Marshall signed an agreement with the local Yalesumni Indians whereby the whites leased some 12 square miles for 20 years in exchange for $150 a year worth of tools and clothing.[7]

Governor Mason, at Monterey, when informed and asked to ratify the agreement, rejected it out of hand.

The news trickled out, from the sawmill in the Coloma valley – where the labourers at first had agreed to keep the secret – to Sutter's fort, to Monterey, and on to San Francisco. It was on 15 March 1848 that the story was first revealed in print, in a short article of the San Francisco *Californian*, the first – and very small – newspaper in what was then, with a mere 850 inhabitants, still a very small city. The great awakening was about to come. The nearby foothill region was probably the greatest trove of placer gold on earth.[8]

Erosion, caused by the weather, had, through the millennia, broken the gold first out of narrow veins in the rocks into boulders and then into pebbles and finally into even smaller pieces that were washed into the rivers and creeks of the western slopes of the Sierra Nevada.

More stories and evidence in the form of samples reached San Francisco during April and early May. The spark suddenly exploded on about 12 May. Within three days, half of the entire population of San Francisco set out for the claims at the American river. The rush was on.

By July, some 2000 white men and as many Indians were working several forks of the American river and other nearby streams. Each man's take was between $30 and $50 a day. Some – admittedly very few – made as much as $10 000 in a week. By the end of July, gold worth some $4 million had been taken out.

By 15 August, another edition of the *Californian* came out – the editor had himself tried his luck by then. He now estimated that some 4000 miners were at the diggings, and that most of them were clearing $30–$40 per day. And inflation had already reached the area – the prices of most items had risen four or five times.

Most, if not all, of the 1848 gold rushers 'made it'; it was difficult to fail at the beginning. All of the creeks and rivers which were tributaries of the Sacramento river in the north and of the San Joaquin river in the south contained gold. The greatest fear of the early miners was not that there would be too little gold, but that there would be too much and that the price would go down. And yet, with so much success all around, Sutter and Marshall failed. The rich claims eluded them.

River mining in the California diggings.
Source: Louis Simonin, *La Vie Souterrane*
(1867).

By the early autumn of 1848, the first outsiders arrived, from Hawaii and Oregon. Army and navy men deserted, as did sailors from ships that reached San Francisco. The first Mexicans came from Sonora, and by October the first shiploads of Chileans arrived from Valparaiso.

In those years, when the Panama Canal was still 50 years away and cross-country treks were quite tedious, many travellers to and from California, and especially those with money, went by ship around South America. En route, Valparaiso was a much frequented way-station.

Early in December, President Polk mentioned the gold finds in his message to Congress. Messengers from the governor and the military authorities in California now also reached the east and samples were put on display at the War Department. The newspapers competed with each other for ever fancier stories.

The then (US) military governor of San Francisco sent an aide, Lt Lucien Loeser, with a letter and 230 ounces of gold (6.3 kilograms, not exactly a very light load, that had to be carried in his hand all the time) to Washington. Loeser took a sailing ship from San Francisco to Peru (because that is where, presumably, the easy winds would take the ship). From there, he took a steamer back to Panama where he crossed the isthmus by mule and canoe.

Once on the Caribbean side, he took another steamer to New Orleans and then proceeded by stage coach to Washington. It took him all of five months. The gold he brought was the sample then put on display at the War Department.

By January 1849, the rush became a flood. Groups formed all over the eastern states to join efforts and resources; many became regular companies, with articles of association and membership fees.

News of the finds had now also reached Europe and Australia and everybody wanted a share. Between 16 and 12 January 1849, 23 ships with about 1500 passengers left eastern and southern US ports for California.

The story of 1849 was a very different one from that of 1848. Whereas in the first year the gold rushers had been few – compared with later years – and the pickings rich, by 1849 the richest bonanzas were gone. Life was now tough. Many if not most of the easterners and foreigners who now arrived were not used to the hardship of the mines. Adventurers, gamblers, hoodlums, prostitutes and plain bandits also came. Racial problems arose. Mexicans, Chileans and other Latins were suddenly disliked and picked on. Throughout 1849 trek after trek of those who had chosen the transcontinental overland route also arrived – from Arkansas via Santa Fé and then through southern Arizona, from Missouri and Iowa via Kansas, Nebraska, Wyoming and Utah to Nevada and on to California, and also via the old Oregon Trail or, at least partially, into northern California. Some 25 000–30 000 seem to have taken these trials in 1849, and between 50 000 and 75 000 in 1850.

By late 1850, the actual gold rush was over. The pickings were ever smaller, and more and more of the rushers were disappointed. Some returned and spread dismal stories. Yet new ones kept coming, partly because the momentum was hard to stop. In 1851, another 15 000 reached California via Panama, but the overland flow declined to just a few thousand. Nevertheless, the overall value of the gold extracted kept rising – from an estimated 7 tonnes in 1848, to 20 tonnes in 1849, to 85 tonnes in 1850, and to a peak of 120 tonnes in 1853.[9] No mean production!

A real revolution now came, or better still, the actual gold mining now began, with heavy machinery and large investments.

The town of Sacramento had sprung up in 1849. And San Francisco had grown from 850 people early in 1848 to some 30 000 in late 1850, by which time the entire population of California was estimated at at least 100 000, but probably closer to 150 000.[10]

It was also in August of 1850 that Congress approved statehood for California; President Fillmore signed it into law in September.

The gold discovery in the foothills just east of San Francisco, and the rush for it in 1848–50, made America aware of California, brought the entire west closer and, in a way, was the first substantial step towards a unified United States.

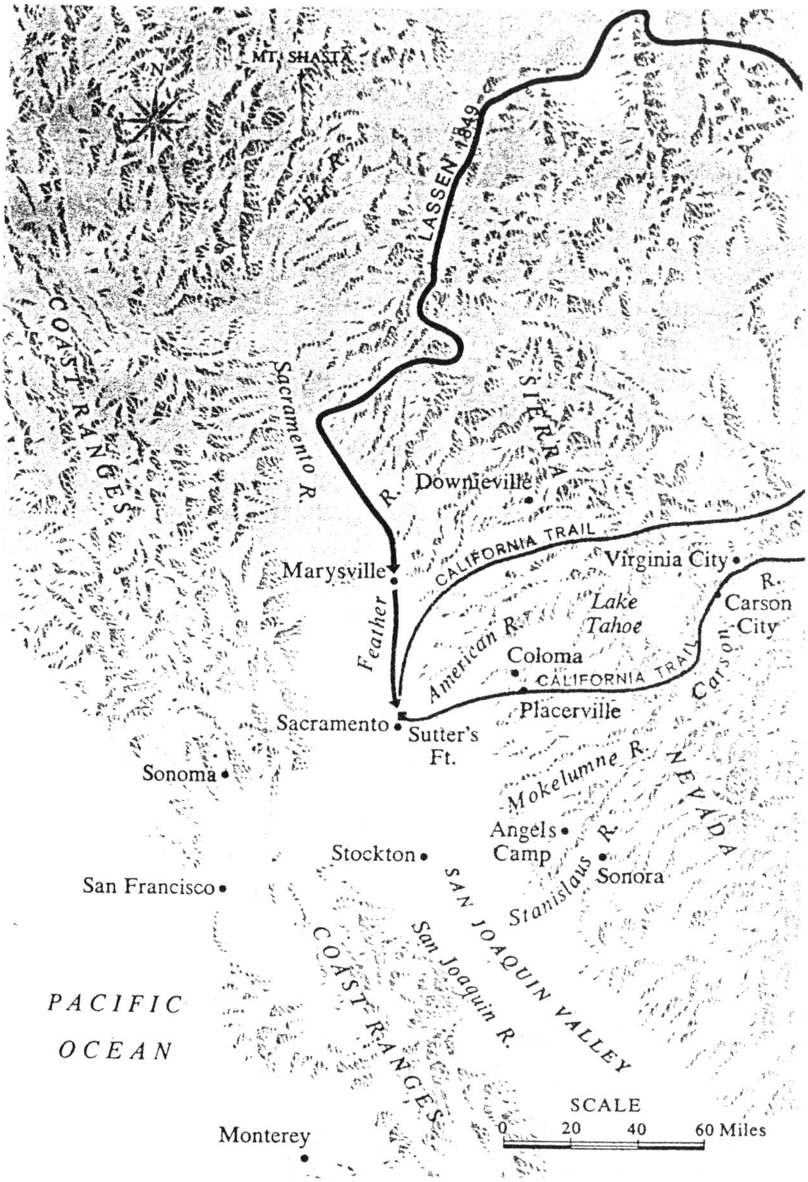

The California gold rush 1848–51: the overland push towards the diggings.

The California gold rush must be seen in its broader social, political and economic contexts. Between 1845 and 1848, immediately preceding the rush, the person who has come to be considered one of the ten best US presidents, James K Polk, greatly influenced the destiny of the young United States with three fundamental steps:

1. He approved and implemented the annexation of Texas.
2. He fixed the northern boundary of the then Oregon Territory along the 49th parallel.
3. He made war against Mexico which led to the annexation of the northern part of Mexico, i.e. the modern states of New Mexico, Nevada, Arizona and California.

These events, and the gold rush of 1848–50, as well as the resulting settlement of the West Coast, essentially rounded out the territory of the United States the way we know the country today.

Many were the fabulous stories surrounding the gold rush. Like $300 worth of supplies becoming worth $1000 in less than three months, or a pound of coffee rising from $1 to $4 in the same period, in 1848. Or a breakfast at Sutter's mill, of sardines, bread, cheese and beer, costing $43. Or the escalating price of iron pans, to pan gold, which cost 20¢ each at the beginning of the rush and later went for $8–16 a piece when one quick thinker bought up all he could get. With such profit potential, who needed to pan gold? Staying in town and selling goods must have been as profitable and often more so!

By 1849, almost half of the population of California, then about 100 000 inhabitants, consisted of foreigners. It included many from Chile, Peru, Hawaii, China, Japan and even Tahiti. Chauvinism, and exaggerated nationalism became important. The 'Americans' (few indeed were those who could claim not to have any foreign blood at all in their veins, and those that could, did not count – the natives, or Indians, as they called them) did not so much object to foreigners *per se* as for reasons of power, envy and money.

A 'Foreign Miners' Tax' of $20 per month was assessed. It was aimed at Mexicans and Chinese, but affected all foreigners. In 1851, the tax was reduced to $4 a month.

And what of the natives? In the first year, 1848, they were welcome as workers. In 1849, they were driven away from the gold fields, and by 1850 they were driven into the high sierras, most to starve there.

And what of women? In 1850, only 7% of California's population were women, and in the mines only 2%. Mark Twain wrote they were 'virtuous according to their lights, although their lights were somewhat dim'. In December 1850, President Louis Napoleon of France instituted a lottery to send the poor to California. Many of the women chosen to go were whores.

By 1860, there were 380 000 people in California.

The Comstock Lode

Gold was the craze of the 1850s. Hardly had the Sierra Nevada foothills ceased to be of interest, than stories of spectacular discoveries of gold in a creek in New South Wales, Australia, caused a new stampede, in 1851 and 1852. In 1858, it was the turn of the Fraser river country in British Columbia and Cherry creek in Colorado. But the biggest of the discoveries only came in 1859 – the gold and silver find of the Comstock Lode at what came to be called Virginia City in Nevada. The ensuing stampede almost emptied the California mines.[11]

Former US Senator RL Neuburger has described the Comstock Lode as probably the richest single piece of mining property yet discovered.[12] It is said to have yielded throughout its history some $600 million in gold and some $500 million in silver – at the prices prevailing then. In 1877, its output was some $37 million. Its richness probably contributed very substantially to the financing of the Civil War. And its opulence probably also contributed greatly to the phenomenal industrial spurt of the United States after the war. And, once more, a state, Nevada, came about because of a mining discovery, and what was in about 1850 almost totally empty country, had, by 1880, some 62 000 people and a thriving mining industry.

The story starts back in 1850. What was to become the Comstock Lode was part of a mountain right off the Carson river valley. The Sun Mountain, highest of the Washoe range, towers some 8000 feet above sea level and some 5000 feet above the floor of the desert. Alongside it ran the immigrant road from Fort Laramie and Fort Bridget via the Great Salt Lake desert, up the Humboldt and Carson rivers past Lake Tahoe and across the Sierra Nevada into the California gold country. Between 1849 and 1851, some 60 000, possibly as many as 150 000, of those bound for the California gold diggings had gone past the Sun Mountain. Among them, early in 1850, was a young Mormon by the name of William Prowse. When his wagon train was temporarily stopped at the mouth of a deep canyon near Spafford Hall's Station on the Carson river because of snow in the pass higher up, Prowse apparently took a milkpan, went into the canyon, picked up some dry creek sand, placed it in the pan, whirled it rapidly in the wind – and was left with some gold flakes.[13] But he and his group were on their way to the real Eldorado in California and nothing could deter them from that objective. They called the gulch 'Gold Canyon', and so it has remained. John Orr, the boss of the wagon train, returned the next day and found a nugget the size of a gold coin. But why waste time with small pickings, when riches were just beyond the mountain range? On they went! Others who passed through Spafford Hall's Station heard of the find and a few tried their luck. One digger washed out $30 000 in a short time; another washed out 600 ounces in one day. But all complained that the ore was too flaky and that it was 'alloyed with some fake blue stuff that chokes the rockers and carries away the quicksilver'.[14]

Early in 1851, a new group of Mormons passed by, including a teamster by the name of James Finney. His name has been corrupted into Virginny, or maybe it was because he came from Virginia. His real name was James Fennimore, but that had best be forgotten because of a knifing – and a dead man – over in California. He started to wash for gold in the canyon and to wash his thirst down with 'tarantula juice' at Hall's Station. The Mormons settled in the valley and brought in 100 Chinese to build an irrigation ditch to their farms. Spafford Hall's Station soon became Chinatown.

Other types similar to Fennimore also went into the canyon to work. Among them was one Bill Thorrington. He was called 'Lucky Bill' and he made good. He soon set up a station and began to levy toll charges on the valley road. He called it the Carson Canyon Toll Road – and it was with guns that control was exercised and tolls collected.

And then a Canadian sheep-herder by the name of Henry Thomas Paige Comstock appeared on the scene. A town now sprang up, Johntown, some four miles up the canyon from Hall's Station.

The further up the canyon they worked, the worse the blue stuff got. One day a Mexican miner examined it and exclaimed loudly, *bueno*, and he waved his arm towards the mountain and told them: *mucha plata, mucha buena plata*. They took it as a confirmation that high up there was much gold.

Among yet more newcomers were Allen and Hosea Grosch, sons of a Universalist clergyman from Utica, New York. They were fine, upstanding young men, contrary to most of the riff-raff so far at Sun Mountain – yet they too had come west to 'make their pile'. They even knew a bit about mining and had brought some technical books along. They went further up the canyon than anyone had so far and began to work with a rocker. The 'blue staff' bothered them also, but they did not leave it at that. They ground it into a fine powder and then baked it in a small oven. They got a small button of metal in the crucible. The next test was vital. The button dissolved in nitric acid: the 'blue stuff' was silver.

They decided not to tell the Johntowners, and they really stuck to it. For the next few years they prospected up and down the canyon during the summers, and during the winters they worked at odd jobs in California to earn money for further prospecting. It sounds odd but is true. One would think they would have worked for gold right there in the canyon rather than working at odd jobs elsewhere.

By late 1855, they had worked their way up to the divide and they finally found a thick, dark blue vein which sparkled with gold. They now decided to organize a company and bring in machinery. But first, they were to spend one more summer prospecting and assaying. Exploratory assays gave values of \$3500 per tonne.

But then their resources ran out. It was August 1857. By now they had taken in as a partner George Brown, station-master at Gravell Ford, on the Humboldt river, who offered to give them all his savings, some $600. Suddenly, disaster struck. Hosea cut his foot with his pick. Blood poisoning soon set in and a few days later he was dead. News also came in that bandits had attacked Brown, had taken his $600 and had murdered him.

Allen was now all alone, although one Laura Ellis – a hotel cum restaurant cum laundry operator in Johntown – had offered the Grosches her savings and had been promised a share.

Allen Grosch decided to depart once more, to spend the winter in California. Before leaving he called in Comstock, to tell him of his claims – but not of the 'blue stuff' – and to ask him to stay at and protect his cabin and the notebooks and instruments until the spring, in return for which he would give him a one-fourth interest in the claims. Comstock agreed.

Allen Grosch departed and Comstock settled in for the winter.

But Grosch had waited too long. The full force of winter caught him on the western side of the pass. Almost within sight of some diggings he passed out. When the miners picked him up, both his legs were frozen. A few days later he was dead. The secret of the Sun Mountain was, apparently, lost.

But there was still Comstock. He was wintering with the instruments and notebooks in Allen Grosch's cabin. When he heard of Allen's death, he decided to open the boxes. The samples, including 'blue stuff', meant nothing and the papers were unintelligible. Comstock got furious. He busted the equipment and threw the papers into the fireplace. The last piece of paper he gave to the flames was his contract with the Grosches. He was now not any more bound by the past; all he had to do was find the claims of the Grosches, register and exploit them. And he would be as rich as Croesus.

So he began to look for the claims, but 'Old Virginny', James Fennimore, was right on his heels.

Comstock had samples of likely outcroppings assayed at nearby Grass Valley, but before the assayer could give his report, the latter was murdered. A geologist declared the ore to be common blue limestone. Finally, an assayer in Sacramento found $2200 per tonne worth of gold, but silver was not mentioned.

It was in January of 1859 that Old Virginny and some buddies found 'blue stuff' just sparkling with gold at what came to be called Gold Hill Outcroppings. They staked out claims of 50 feet each; Comstock came late and took what was left – one of the poorer claims. By next morning, most of the Johntowners had staked out claims.

The claims were very rich. They gave $300 to $500 every few hours. Still they cursed the blue stuff. Two Irishmen named Pete O'Riley and Pat McLaughlin – apparently of the original band of rogues who had settled in the canyon – located another rich vein at a place called Spanish Ravine. Comstock forced them to cut him and two partners, including Old Virginny, in – each for a fifth of what was in total a 100 foot claim. Then Comstock bought Old Virginny's part for a half blind horse and a bottle of whiskey. Old Virginny's 20 feet came to be worth, a year later, some $60 000 – an expensive horse.

Within a week, Comstock incorporated Penrod, Comstock & Co and called his part of the mine 'Ophir'. Yet he kept referring to his and the nearby claims as Comstock's Lode. And by that name it entered history.

Not so the town. It became known as Virginia City, because Old Virginny one day stumbled over a boulder and a bottle of whiskey he carried fell and broke and the liquid disappeared into the ground. Virginny, nonplussed, got up, stood to attention and proclaimed rather formally that he was baptizing the place 'Virginia Town'. Soon it was Virginia City, and thus it is still known today, even though now it is no more than a ghost town.

In June 1859, a man called Stone asked Comstock about the 'blue stuff' but only got the standard answer that it was worthless and that it only made the mining of the gold more difficult. But Stone sent samples over to an assayer in Grass Valley, who now found 'almost solid silver' in assay after assay.

And the silver rush was on.

In October of 1859, a pack-train of 80 mules with tonnes and tonnes of silver ore reached the Mosheimer furnaces in San Francisco. The ore was worth $5000 per tonne. What more proof of the wealth of the lode was needed? Nevertheless, Comstock still thought it was only a pocket. Thus, when in August of 1859 Judge Walsh and Joe Woodworth offered Comstock $5000 for his share in the Ophir, he still refused it saying that the Comstock was worth millions, but when they offered him $11 000, he sold out – a share worth millions for a pittance. Exit Comstock, except for the name of the lode, which stuck. For a while he bragged about how he had fooled 'that California rock shark', but a few years later, when nobody listened any more to his fancy tales, he pulled the trigger on a Colt and blew out his brains.

A George Hearst now appeared on the scene. He had little money for a meaningful stake, but much conviction and courage. He borrowed $1000 from a hotel keeper and bought McLaughlin's interest in the Ophir for $2500, and set to work.

Soon Hearst and two partners, Head and Atwood, took 45 tonnes of blue stuff to a San Francisco smelter on muleback and saw it give $80 000. In a few years, Hearst was one of the first Washoe millionaires – and the *San Francisco Examiner* and a newspaper empire followed.

Meanwhile Pat McLaughlin took a job as a cook in a sheep-herding camp at $40 a month. Pete O'Riley had sold his share for $40 000, which he then invested in a hotel in Virginia City. He then lost that, in turn, by boring a tunnel through an oreless mountain. He died penniless in a lunatic asylum.

Soon the Rothschilds had their representative, one Donald Davidson, up at

Virginia City. He secured an option on one of the mines and some experts came over from Europe. But they proclaimed it a 'worthless contact vein'. Five years later, at much greater cost, the Rothschilds were after the same properties.

By late 1859, Wells Fargo & Co had set up the first bank in Virginia City, and Adolph Sutro, a Jewish cigar dealer from San Francisco, kept talking about tunnelling into Sun Mountain from way below, at the level of the river, to mine the mountain 'inside up'. Everybody said he was a lunatic.

A telegraph line was laid, actually hung from tree to tree, all the way to Placerville – until the next storm tore the wires down. They were restrung the next spring. By then, in April of 1860, the Pony Express had also reached Virginia City.

For a while, in 1860, Piute Indians threatened the city and the works. Some skirmishes and even a battle followed, but then everything quietened down again, for a while. The coming confrontation between the South and the North also threw its shadow. Loyalties were divided as to the origin of each individual. Then, in June of 1860, the Federal government, mindful of the importance of the route to California, set up Fort Churchill on the Carson river, near Virginia City. And a newspaper came, the *Territorial Enterprise*, at which, later on, Mark Twain was to work for a while.

The shaft of the Ophir sank deeper and deeper and the ore got richer and richer. At first, the miners worked it with a windlass and a bucket to lower men and hoist out the ore. Only ore worth $1000 a tonne was shipped out – to England, in bags. Lower grade ore was set aside for the future. Ore worth only $50 a tonne or less was used to pave the streets of the city.

The miners next used horse and whim, and for a while production continued well. Then water accumulated down in the mine, at the 50 foot level; they installed a 15 horsepower steam engine to pump the water out. It was then the most powerful engine in the west. Yet the final answer was an adit. The tunnel went through 1100 feet of hard rock and – four months later and after a cost of $10 000 – reached the shaft line 155 feet below the surface. At the 175 foot level, the vein broadened to 65 feet! How could so much wealth be all in one place?

They now also tackled the question of milling. Several methods were tried, first and foremost the old patio process of the Spaniards. It consisted of letting horses and mules trample again and again over the ore mixed with water and salt until crushed to paste. But the quantities were just too large and the animals suffered too much. They eventually settled for large iron pans with huge stamps lifted by steam pistons and then falling on, and crushing, the ore. The process was in time refined and people came from all over the world to study it. Amalgamation, with mercury from nearby California, salt and some copper sulphate, followed. To bring in the salt from across the desert, somebody imported camels, which could carry loads of up to 800 lb.

After the Civil War broke out, a group of southerners, some 200 of them under one Dr McMeans, raised the flag of the Confederacy and took control of Virginia City. But it was only a few hours before Union troops from Fort Churchill recaptured the city and the lode.

In 1861, Congress made Washoe into the territory of Nevada. Virginia City and the Sun Mountain were really so far all there was to Nevada. James W Nye was appointed governor.

Many and varied are the stories about the Sun Mountain in the 1860s – stories about the gunmen, the whores, the saloons and gambling joints; about the unbelievably primitive conditions underground, the cave-ins, noxious gases, water, fire; about the 10 storey down 'building' beneath the top – but none is as fabulous as the statistics of the sheer wealth that poured from the mountain. One of the mine managers made $40 000 a year. One of the mine owners on his wedding night used his water tank to dispense champagne. Some of the horses were shod with silver shoes.

By 1864, some 25 000 people were working on the Comstock Lode, 9000 more in Gold Canyon. By then, Virginia City was, next to San Francisco, the most important city on the West Coast. But by then the richest part of the Comstock Lode had already been mined out and most of the mines were in litigation. Nobody had taken care, at the beginning, to measure the claims accurately, and no surveyors had been around. Judges and justice were corrupt and for hire. The shafts had also by now been sunk as far down as was possible, and mining costs had risen substantially. Some of the mines began to close. Attempts were made to cut wages from four dollars a day to three and a half, but the attempts were dropped when several of the mine managers were almost lynched.

By 1865, the rush and the boom were over, most of the mines abandoned, drowned or caved-in. And quickly the miners drifted on to other workings elsewhere. It would take other equipment and methods to keep working the Sun Mountain.

The Guggenheims

The next important chapter in American mining is not the story of a mine, or of a region, but of a family: the Guggenheims. For some 50 years, between 1881 and 1930, they dominated the American mining scene like nobody before nor after them.

Part of their story has already been told in the chapters about Mexico and Chile (see pages 38–9 and 145).

The Guggenheims came from Lengnau, in Switzerland. In 1847, Simon Meyer Guggenheim, his second wife Rachel, and their 12 children, among them Simon's only son Meyer, took off for American and settled in Philadelphia. Life in Switzerland had been quite difficult for Simon as a tailor, and for Meyer as a pedlar. They had not been able to choose a profession freely, had not been allowed to own land, had not been able to live where they wished and had even encountered difficulties in marrying.

Both Guggenheims began by peddling – shoe-strings, lace, pins, needles and other small and easy to carry goods.[15] Meyer's bestseller became stove polish. To increase his margin he quickly learned how to make it himself. Three years later, in 1852, Meyer had progressed sufficiently to be able to marry his stepsister, Barbara. Meyer and Barbara had 11 children: Isaac, Daniel, Murry, Salomon R, Jeanette, Benjamin, Simon, Robert, William, Rose and Cora. The sons were to form a dynasty unique in American mining history.

By 1873, Meyer Guggenheim, at age 45, had progressed from peddling and stove polish manufacturing and had become a prosperous wholesale spice merchant. He next branched out into the making of lye, a business he sold a few years later for $150 000. He next acted on a tip to buy shares in a small midwestern railroad, the Hannibal and St Joseph Railway, which serviced Kansas City. Meyer picked up 2000 shares. Soon Jay Gould, the speculator and railroad tycoon, trying to put together the Missouri Pacific Railroad, wanted the shares, and Meyer held on to them until he received $300 000 for them.

Meyer now had close to half a million dollars. He decided a second fortune could be made by importing lace and embroidery from Europe. And once more he prospered. In 1877 he set up the firm M Guggenheim's Sons – incorporated in 1882 – in which each of his seven sons had an equal share.

In 1881, Meyer Guggenheim took a real plunge. He bought a one-third interest in two lead and silver mines near Leadville, Colorado. An acquaintance, a grocery store owner, had bought the mines with borrowed money, and when production costs turned out to be expensive and the results only mediocre, he needed help.

Meyer went to inspect the investment. Leadville was then a 'roaring' western mining town, with 120 saloons, 150 gambling houses, 1 opera house, and 35 whore houses. And the mines? They were flooded. It was to cost $25 000 to dewater them. Meyer's partner was unwilling to invest more and Meyer quickly acquired control of the properties. But it was all really haphazard and accidental; Meyer knew nothing of mining. He acted by instinct.

He returned to Philadelphia. More money had to be sent, and soon he had invested more than $50 000. But suddenly luck turned his way. A cable came: 'RICH STRIKE FIFTEEN OUNCES SILVER SIXTY PERCENT LEAD'. Now he owned the mine with the highest silver content in the Leadville district. Soon Meyer's profit came to $17 000 per month – on a total investment of slightly above $50 000. By 1887, the mines had produced 9 million ounces of silver (about 270 tonnes) and 86 000 tonnes of lead. In the same year, Meyer earned $750 000 from the mines, and by the time they were exhausted, they had yielded him $15 million.

The ore was shipped to the Holden smelter in Denver, Colorado, for refining, but the smelter, to Meyer's view, was taking too much of a cut. He negotiated with Holden and they pooled their assets, forming a new company, the Philadelphia Smelting & Refining Company. Holden owned 49%, Meyer Guggenheim 51%. The next step was to build a new smelter, at Pueblo, Colorado, for $500 000.

By now, son Benjamin was the bookkeeper of the mines in Leadville and son William worked in the assay department. The older boys, Isaac, Daniel, Murry and Salmon R were still in the embroidery business, both in the States and in Europe. But they soon joined forces in the mining and smelting part – especially as the smelter suddenly ran into difficulties. Costs increased, strikes paralysed work, prices declined. They were losing close to $100 000 a month. As a result of the poor performance, the partners now wanted out and M Guggenheim's Sons became the sole owner of the Pueblo smelter. Conditions finally improved, the strikes diminished and the price of silver stabilized.

The Guggenheims now moved to Now York, the real centre of the country. Meyer retired 'officially' but remained active. Daniel became the leader of the family. Issac took on financing and accounts; Murry handled sales and Salomon R took charge of public relations. Benjamin ran the show out west, Simon bought ores and helped in public relations with the miners, and William worked at the smelter.

The government helped. The Sherman Silver Purchase Act 1890 drove the price up and immediately contributed to substantially improved profits.

Success pushed the Guggenheims into other ventures. Expansion into mining in Mexico was a logical step. Labour costs south of the border were much lower. They imported ore across the border for treatment at the Pueblo smelter. But then the McKinley Tariff Act burdened imported ore with a substantial duty, instigated by other American mine owners who wanted to keep the cheap foreign ores out to obtain a better price for their own. But the Guggenheims were not to be caught napping. Daniel and Murry investigated conditions in Mexico, and, following meetings with Mexican President Porfirio Diaz, the Guggenheims obtained concessions to build smelters at Monterrey and Aguascalientes. And all the equipment could enter duty-free, as was described in Chapter 3 (page 39).

About this time, Meyer told his sons they ought to strive to obtain control of all mining an smelting in North America.

By 1895, the three Guggenheim smelters gave a net profit of $1 million a year.[16]

In 1889, they formed Guggenex – Guggenheim Exploration Company – a public enterprise with outside shareholders such as King Edward VII of England, to search worldwide and develop promising mines. At about the same time, Henry H Rogers, William Rockefeller and the Lewisohn brothers formed first the United Metals Selling Company and then the American Smelting & Refining Company, later called ASARCO, with the clear objective of controlling all non-ferrous mining and smelting activities in North America. The Guggenheims were asked to join the trust but declined; they preferred to continue as a closely knit family group. The stage for a major battle was now set. The ASARCO group was out to ruin the Guggenheims because wasteful competition had to be eliminated.

The Guggenheims realized that they were at a tremendous disadvantage against ASARCO, which combined the might of 23 smelting companies. But the Guggenheims decided to fight the giant, apart from two of the younger brothers, Benjamin and William, who wanted out. They did not have the stomach for the major battle now coming.

In 1900, the workers at ASARCO struck for two months. During that time, the Guggenheims bought up cheaply those ores that could not be smelted by the ASARCO trust and then dumped the metal on to the market at low prices. By the end of 1900, ASARCO's profit was $3.6 million and the Guggenheim profit was $3.5 million, but the Guggenheims were only a quarter as big as ASARCO. ASARCO shares began to fall, and the Guggenheims bought.

Major power plays followed and suddenly the Guggenheims were on top of ASARCO. Daniel was chairman of the board and president, Salomon was treasurer, and Isaac, Murry and Simon were also on the board. They controlled 51% of the company.[17]

With the boys now definitely established as a power in world mining, Meyer Guggenheim's life dream had come true. He died in March of 1905 a happy man.

ASARCO, M Guggenheim's Sons (after 1906 called Guggenheim Brothers) and Guggenex now really expanded beyond the frontiers. In 1907, they teamed up with J P Morgan and Jacob Schiff to go after one of the richest copper deposits on earth – on Alaska's Kennecott creek. Some of the samples assayed 70 and 75% copper.

A $25 million, 300 kilometre railroad had to be built and a harbour laid out, as well as a mining camp and warehouses. The railroad was completed in 1911, by which time the mine was already producing ore assaying 75% copper which went down to the Guggenheim smelter in Tacoma, Washington, for processing. Thereafter, the profits just kept rolling in.

Other ventures of the Guggenheims included a refinery at Perth Amboy, a mine at San Luis Potosí, a smelter at Veracruz and a lead refinery at Salt Lake City.[18] They also went into the Congo and into Angola.

And they also tackled the Klondike, with giant dredges, to pick up industrially what the individual gold-washers had left behind. But the Yukon Gold Company was only a partial success, and another venture, Nipissing, near Hudson Bay, was a complete failure.

However, Bingham Canyon in Utah was another huge success. This time it was low grade copper ore, only 2% copper, but enormous quantities of it. It was a porphyric ore body, with the copper disseminated, not concentrated in veins. They applied new technology, open pit mining, with huge steam shovels.

By 1910, Bingham Canyon paid a $4.6 million dividend. By 1935, the dividends paid topped $200 million. In 1990, Bingham Canyon was still a profitable producer.

By 1910 – probably the high point of the Guggenheims' involvement in US and world mining – their interests also included the Chuquicamata mine and Braden Copper in Chile, as well as innumerable smaller American mining ventures. When World War I began, the Guggenheims controlled 80% of the world's copper, silver and lead, and were in a position to dictate the prices of all three.[19]

World War I found them thus ideally positioned to cash in on at least two metals essential for any war effort, and they benefited. By the end of the war, their copper properties alone had paid out $210 million in dividends.[20] They were criticized as profiteers; when asked to reduce their prices they refused, and only when President Wilson threatened expropriation did they agree to peg the price at 23.5¢/lb – still more than double the pre-war price.

The end of the war thus found the Guggenheims vastly richer and more powerful than at its outbreak; but also with a somewhat tarnished reputation – for profiteering and poor labour relations. They were probably also by then the second richest – next to be Rothschilds – Jewish family in the world.[21]

The year 1916 had been a milestone in the history of the family. William, who in 1901 had bowed out of the partnership, now suddenly felt entitled to a share in the Chile Copper Co, which was just beginning to operate Chuquicamata. When the active partners refused to cut him in, he sued his brothers for $10 million, claiming that his earlier waiver to participate was based on deliberately false information as to the mine's potential. The dispute was settled out of court, with William probably collecting close to $5 million.

In 1922, the board of directors of ASARCO accused the Guggenheims of having milked the company by burdening it with undue risks and expenses. The family promptly lost control of the board. Their decline had begun. Isaac died the same year.

In 1923, the Anaconda Copper Corporation offered to buy 55% of Chile Copper Co, the owners of Chuquicamata, for $70 million cash. After much squabbling inside the family group, they sold.

Daniel Guggenheim was now 67 years old, Murry 65, Salomon 62 and Simon 56. Each was worth about $50 million. Isaac and Benjamin were dead, and William, following the lawsuit, totally alienated.[22] From now on, almost all of them began to enjoy and spend their money, several staying in close touch with mining, and even risking new ventures once or twice more, as in 1931, into Chilean nitrates, which turned into a colossal failure in which Murry, Salomon and Simon lost some $60 million. It was only during World

War II that the family recouped most of its losses, partly through a sudden resurgence of demand for Chilean nitrates.

Anaconda

Where the previous story was one of a family and many mines, Anaconda is the story of a mine with many different personalities. Anaconda is also a continuation of those lusty tales of the California gold rush and of the Comstock Lode.

In 1875, one Michael Hickey, a former soldier of the Union army, staked out a claim on a hill in Montana. At the foot of that hill lies today the city of Butte. The claim became the Anaconda mine, and out of the mine grew one of the largest non-ferrous mining and metal-fabricating organizations in the world. The early history of Anaconda is also the story of Marcus Daly, an Irish immigrant boy who worked as a stevedore in New York, Cornelius F Kelley, who worked as a water boy to a gang of railroad construction workers for a dollar and a half a day, John C Ryan, who at first clerked in a country store, and Fritz August Heinze (introduced in Chapter 13), one of the more unsavoury characters of the Old West.

But the story of Butte and Montana began much earlier. When the California diggings were worked out, most of the miners moved on, to Utah, to Nevada and to British Columbia. Some also went to the Idaho country, of which Montana was then a part. They were after gold, and maybe silver, although, as we saw in the story of the Comstock Lode, it took them a while to figure out what it looked like. Copper, the metal which in time was to make Butte and Montana famous, was little appreciated. It was not wanted, and hardly needed, in the west. Its value was comparatively low and transportation costs high. Gold and silver paid, copper did not.

What Hickey staked out in 1875 was a silver claim. And the town of Butte was a gold and silver camp.

The first gold find in Montana dated back to 1852; the first major strike was at Alder Gulch in 1862. Some 30 million dollars worth of gold was taken out in three years.[23] A town sprang up and they called it Virginia City, after the place in Nevada made famous by the Comstock Lode. In the late 1860s, Virginia City abounded with 'gold, Indians, whiskey and fugitive rebels'. Another discovery was called 'Last Chance Gulch' and it became the city of Helena.

The original operation in Butte was begun in 1858 by William Allison and G O Humphreys. They sank a shaft at what they called the Original Lode and hauled gold bearing ore to the Silver Bow creek for washing. After a few weeks they were down to about 80 feet – and found more and more copper. The red metal was a great nuisance as separation was difficult. Yet, in 1866, a first shipment of Butte copper–silver ore went to the newest smelter, at

Swansea in Wales. Soon, however, one Dennis Leary built a local smelter. It began operations in 1867. In the same year, a stamp mill was set up, as well as additional furnaces to treat copper and lead ore. In 1868, another mill went up and the town site was laid out.

But silver still continued to be the dominant metal. Production increased steadily. By 1875, Butte had 4000 inhabitants; the first silver bullion was produced by amalgamation and roasting. Butte was suddenly the leading silver-producing district in the world.[24]

Marcus Daly now arrived in Butte. He had been hired by the Walker brothers, bankers of Salt Lake City. Daly had prospected before at the Comstock Lode, as well as at the rich Ontario and Ophir mines, and was considered something of an expert in silver mining. He was now to advise the Walkers on which silver mines in Butte to buy. In 1876, he recommended the Walkers purchase the Alice silver mine for $25 000; Daly stayed on to work as superintendent.

Michael Hickey had come to Butte in 1866, to prospect for silver. He was only moderately successful and it was only in 1875 that he staked out the Anaconda claim, 1500 by 600 feet. He chose the name because the word had caught his fancy years ago in a newspaper article by Horace Greely.[25] Hickey worked his claim with some success and acquired more claims, so that by 1880 he owned 25 of them. But then he ran out of money and had to take in a Charles Larabie as a 50% partner, in order to sink his shaft deeper. But the infusion of Larabie's money was not enough.

One day Marcus Daly happened to pass by. He had just quit working for the Walkers at the Alice mine and he had some cash. Hickey offered him a one-third share in exchange for deepening the shaft still further. Daly accepted.

After further development work, Daly was so enthused about the property that he immediately obtained from Hickey and Larabie an option to acquire their shares. Daly then offered the option to the Walkers; they declined. He next approached the Haggin–Hearst–Trevis syndicate, which had made good money in the California, Utah and Nevada diggings.

James B Haggin and Lloyd Trevis had formed a law partnership in 1850 in Sacramento. In 1853, George Hearst joined them and they moved to San Francisco. They now began to invest in mining and went from success to success. They combined courage with business acumen and luck. The silver mines of Mineral Hill, Nevada, were their first success, to be followed by the Ontario mine. Then followed the Homestake mine in the Black Hills of South Dakota. And now came Anaconda.

Daly travelled to San Francisco to see Haggin. The members of the syndicate were impressed by Daly's story and his faith in the Butte region. George Hearst went back to Butte with Daly to have a look. He agreed with Daly, and the syndicate acquired the option for $70 000. Daly retained a 25%

stake, and the job of general manager in charge of developing the mine.

Late in 1882 they struck it rich. At 300 feet, a cross-cut encountered a vein of near solid chalcocite. The mine became the richest deposit of copper sulphide ever found. And with it, the large scale copper mining industry of the United States was born.[26] The vein was between 50 and 100 feet wide, and the ore assayed 55% copper.

The ore was smelted at the Swansea smelter. It was transported by ox team to Corinne, Utah, hence by rail to East Coast ports, and then by vessel to England. Later, a route via Portland, Oregon, was used, and thence by sailing vessel to England. Between 1882 and 1884, some 37 000 tonnes of ore were shipped out.

Daly and his partners had boundless, almost blind faith in the Butte region and in the future of copper. Without such faith, large scale investment in copper mining made no sense as, after all, world production of copper metal in 1880 amounted to only 156 000 tonnes.[27] But humankind was at the threshold of fundamental changes.

In 1876, at the Centennial Exposition in Philadelphia, Alexander Graham Bell had first transmitted speech through a copper wire; in 1882 some streets in New York began to be lit up from the first electric generating plant. And in 1880, Edison patented the incandescent lamp and began to experiment with the trolley-car.[28] Still, in the year Anaconda opened up, US copper production reached only 44 000 tonnes.

By 1883, Daly was satisfied that the Anaconda ore could be smelted easily, and the syndicate now set out to construct the first copper smelter, at Warm Springs creek, some 50 kilometres from Butte. The site became known as Anaconda.

The concentrator was completed in 1883. A smelter quickly followed, with a capacity of 5000 tonnes of metal per month. A refinery with a capacity of 1500 tonnes per month followed.

In 1881, the railroad from Ogden reached Butte and thanks to this and to the Union Pacific, Butte and Anaconda were now in rapid and inexpensive contact with most of the country – and the growing copper markets. But a major transportation problem still existed between Butte and Anaconda, where the Montana Union Railroad charged 75¢ per tonne of ore for the transport from the mines to the reduction works. When, in 1892, negotiations for a lower fare failed, the mining companies of the Butte region built a parallel road, and lowered the hauling costs from 75¢ to 25¢ per tonne.

Anaconda expanded. Large timber acreage was acquired, as well as coal deposits.

By 1890, the silver properties of the Butte region suddenly ran into problems. A precipitous drop in the price caused seven smelters to close. Yet, in 1892, the Anaconda group reached a production of 100 million lb (50 000 tonnes) of metal, and with it became the largest copper producer in the world.[29]

It was only in 1891 that the partners incorporated the Anaconda Mining Company, with a capital of $25 million. Until then, Daly had managed and Haggin had acted as trustee for the syndicate. In 1895, the Anaconda Mining Company was replaced by the Anaconda Copper Mining Company. By then, i.e. from 1874 to 1895, the mines of the group had produced 350 000 tonnes of copper, 18.5 million ounces of silver (about 550 tonnes) and some 83 000 ounces of gold (about 2.5 tonnes). Yet not 1 cent of dividend had been declared or paid; the partners had put every penny back into the ground and into ever better and ever larger installations.

In 1891, Hearst died. The year 1896 was significant in that a first dividend was paid – $750 000. In the same year, electrolytic refining was begun, as well as the manufacture of sulphuric acid.

In 1899, Trevis died and Haggin retired as president, to be succeeded by Daly. The same year also marked the acquisition of Anaconda shares by the British Rothschilds.

In 1899, a new personality entered the stage – Henry H Rogers, the head of the Standard Oil Company. Rogers owned some Anaconda stock and represented major interests in the Butte & Boston Consolidated Mining Company and the Boston & Montana Consolidated Mining Company, the latter owned by the Lewisohn brothers Leonard and Adolph. Rogers was a firm believer in consolidation and he first merged the two Boston companies. A merger with the Anaconda group was negotiated next, which resulted in the incorporation of the Amalgamated Copper Company, at first with a capital of $75 million, which was soon, in 1901, increased to $155 million. Rogers was president, Daly vice-president. Haggin bowed out and was paid $15 million for his interest in Anaconda. Daly's interest was estimated at $17 million.

With the Rockefellers, James Stillman of the National City Bank of New York, Robert Bacon of J P Morgan & Co, and others on the board, Amalgamated was now a real giant, but it was only a holding company, not an operator.

Daly passed away in 1900. William Scallon followed him as president. It was Scallon who in 1901 hired a promising young lawyer by the name of Cornelius Francis Kelley, who entered upon a long and distinguished career in Anaconda, capped with the position of chairman which he held from 1940 to 1955. He was generally referred to as Mr Copper.

At the time of Daly's death, another personality important for the Anaconda story appears – John C Ryan. He worked at the Daly Bank, one of Daly's private ventures, in Butte, and upon Daly's demise was chosen president of the bank. In 1904, he joined Anaconda Copper as a director, and a year later became its president. In 1909, when Rogers died, Ryan succeeded him as president of Amalgamated.

One of the more colourful stories of the Anaconda history is the Apex litigation with Heinze. It cost Anaconda about $1 million a year to defend itself.

Fritz August Heinze (a thoroughly German Jewish name) came from Brooklyn, New York. He arrived in Butte in 1889, at age 20, a rare blend of Jewish, Irish and Yankee backgrounds and mixing the worst of each. But he was charming. He began as a $100 a month surveyor for Boston & Montana. He quickly figured out that smelting costs were excessive and decided to go to New York to interest investors. He almost succeeded, with sheer gall and charm.

But then he suddenly inherited $50 000 from his grandmother in Germany. He went to Europe, studied, went on to the Columbia School of Mines, and then returned to Butte. Heinze's declared objective was wealth. His principal, if not only, procedure to achieve this end was litigation. And he litigated.

Claim measuring had not been the most precise in the early mining days. In addition, confusing legislation, as well as corrupt lawyers and judges, gave room for unusual methods and characters.

The ruinous litigation between Heinze and the Amalgamated interests ended only in 1906, when Heinze was paid $10.5 million for all of his interests in the Butte region. In 15 years, Heinze had made it from a $100 a month clerk to millionaire.

As a result of the settlement, 110 lawsuits, involving in total some $70 million, were dismissed. Heinze took his money to New York and set up a bank and a stock brokerage house. Both collapsed as the result of others out-speculating him.

Heinze died in 1914, in Saratoga, New York, at age 45, of cirrhosis of the liver.

In 1910, Benjamin B Thayer, then president of Anaconda, proposed that the shareholders increase the capital from $30 million to $150 million and acquire the assets of the other units that formed Amalgamated Copper by issuing Anaconda stock. The consolidation then took place.

In 1914, Anaconda bought all the assets of the International Smelting & Refining Company through an exchange of shares. Anaconda therewith acquired the Raritan Copper Works in Perth Amboy, NJ, and other assets, including a smelter at Tooele, Utah.

While Anaconda began as what was assumed to be a silver mine, it built its power and wealth on copper. But by 1919, diversification began. A phosphate rock property in Idaho was bought. Others in Montana followed. With it, Anaconda entered the fertilizer field. As part of the special requirements created by World War I, Anaconda in 1914 began to mine for manganese and in 1918 also began to produce ferromanganese.

The end of World War I brought huge surpluses of metals, especially of copper. By 1921, with the excess metal still overhanging the market, the Anaconda mines were shut down for a while. It forcefully drove home to Anaconda's management the need for further diversification – downstream integration, as another age would call it. Ownership of fabricating works that could convert the metal into consumer goods would assure a continuous flow of the primary product, and in the process make for a lowering of costs. The merger with the American Brass Company, the largest fabricator of non-ferrous metals in the world, was the result.

American Brass was the successor to one of the earliest American – and New England – industries, which had begun with Paul Revere's copper mill at Canton, Massachusetts, Israel Holmes, who smuggled British skilled workers in wooden casks on to ships bound for the American colonies, Henry Grilley, and the Port brothers, who set up the brass industry in Waterbury, Connecticut.

The merger of Anaconda and American Brass was consummated in 1922 and represented, up to that time, one of the largest deals in American industrial history.

Anaconda next expanded into fabrication by setting up Anaconda Wire & Cable Company, the result of acquiring a number of smaller units. Anaconda Wire & Cable quickly became the largest wire and cable manufacturer in the United States.

The 1929–33 depression affected Anaconda badly. For a while, copper sold for as little as 4.7¢/lb – when costs just before the depression had been at 13¢. Anaconda eliminated dividends, borrowed money, but did not cut wages. Some of the mines and plants were idled.

In 1933, Ryan died. Thayer followed later the same year.

By late 1936, the copper surplus which had depressed the market and disappeared and the price had risen to 11¢/lb. By March 1937, probably as a result of widespread anxiety about the approaching war, the price had risen further, to 16.775¢/lb, but it soon dropped back to 9.9¢ and was only back at 11¢ when the war broke out in September of 1939.

Throughout World War II, Anaconda once more became a major contributor to the war effort of the Allies. Production of all metals was substantially increased, especially copper and zinc, electrolytic reduction of which began in the early 1930s and, once again, as during World War I, manganese. Uneconomic chrome ore mines in Stillwater County, Montana, owned by Anaconda, were also put into production, at the request of the Metals Reserve Company.

In 1942, Anaconda was asked to take over the management of a magnesium operation in Nevada. A Cleveland company, Basic Refractories, Inc, owned the claims through a subsidiary, Basic Ores, Inc. The know-how came from ICI – Imperial Chemical Industries of England. A new company was set up, Basic Magnesium, Inc, in which the British had 45% and Basic Refractories, Inc, 55%. But Basic Magnesium did not have the manpower to put the project together successfully and the War Production Board Called Anaconda in.

The first pound of magnesium was produced in August of 1942, and peak production rate was achieved in September of 1943, with 120 million lb per year. Most of the magnesium went into incendiary bombs. Its use as an alloy in aeroplane manufacture only came after the war. Late in 1944, production was stopped as oil bombs had proved to be cheaper.

Anaconda's history after World War II is a chequered one. Its North American ventures prospered and expanded but its Chilean properties were expropriated by President Allende in 1970.

In Montana, massive low grade ore bodies, such as the Bokely Pit, the Kelley Mine, the Ryan Shaft, and others, were tackled; they assured an abundant supply of copper for many years to come.

It was in 1951 that Anaconda entered the atomic age by joining forces with the Santa Fé Railroad, on one of whose properties in New Mexico, called Haystack, a Navajo Indian had discovered uranium, back in 1947. Production of yellow cake began in 1953.

In 1955, Anaconda began producing aluminium at Columbia Falls.

The same year, the name was changed to the Anaconda Company.

American Metal Co, Climax Molybdenum and AMAX

Another story of drama, colour, luck and wealth concerns the above three companies – with some of the roots going back to nineteenth century Germany.

The American Metal Company Ltd was set up in 1887, to deal mainly in copper, lead and zinc ores and metal. The owners were three, Metallgesellschaft of Frankfurt, Germany, Henry R Merton & Co of London, England and Ladenburg, Thalmann & Company of New York. H R Merton was a grandson of Philipp Abraham Cohen, who in 1821 had set up in Frankfurt-am-Main, Germany, the predecessor firm of Metallgesellschaft. The London firm was thus 'related'. Close ties existed between the London and Frankfurt firms. Their growth in the 1870s and 1880s coincided with the major expansion of industry in the United States after the Civil War.

By the middle 1880s, Metallgesellschaft and H R Merton & Co had achieved prominence – second perhaps only to Aron Hirsch & Co – in the international metals and ore business and branching out to the United States appeared advisable. Major copper finds in Michigan, Arizona and Montana suddenly profiled the United States as a major copper producer.

The Ladenburg family also came from Frankfurt and in 1880 had set themselves up as bankers in New York. Family ties to Metallgesellschaft existed, as a daughter had married Wilhelm Merton. Metallgesellschaft at first used the New York banking firm of Ladenburg, Thalmann & Company to put there one of their men, one Palmer Budd, to do copper trading, mainly purchasing for export. In 1884, Berthold Hochschild, brother of Zacharias Hochschild – one of the founders of Metallgesellschaft – joined Budd. A few years later, the American Metal Company (AMCo) was set up; Ladenburg,

Thalmann & Company received 10% of the shares free of charge and agreed to stay henceforth out of the metal business. Adolph Ladenburg stayed on as vice-president; Berthold Hochschild was made manager.

American Metal Company soon acquired interests in Balback Smelting & Refining Company in Newark, NJ, and Nichols Copper Company in Laurel Hill, Long Island, New York, as well as in ore-buying activities in Mexico and South America. As a result of friendship and reliable service, AMCo was able to establish close ties with the International Nickel Company, the Orford Copper Company, Société Le Nickel and the Canadian Copper Company.

At the outbreak of World War I, AMCo had about 40 shareholders. Metallgesellschaft owned 49%, some 24% were owned by Henry R Merton & Co and the rest by the directors, among them Berthold Hochschild and Jacob Langloth. Most of AMCo's business prior to World War I was the import of copper ores for American customs smelters and the export of US-made copper metal to Europe.

World War I brought substantial changes to the international metal business. Henry R Merton & Co, as German-conducted, suffered severely. With America's entry into the war on the side of the Allies, Metallgesellschaft's share in AMCo was expropriated. By 1920, the US government sold the shares to, among others, Cerro de Pasco Corporation, and the Barney group, as well as the Hochschild family and other AMCo directors. With the sale ended the direct relationship with Metallgesellschaft.

Among the new shareholders was also one Ludwig Vogelstein, whose firm, L Vogelstein & Co had been, prior to the war, the New York associate of Aron Hirsch & Co of Halberstadt, Germany. L Vogelstein now merged into AMCo, and the United States Metals Refining Company, with its plant at Cataret, NY, and controlled by Vogelstein, became an AMCo subsidiary.

It was during the war, in 1916, that Max Schott, then in charge of AMCo's office in Denver, Colorado, heard of molybdenum deposits on nearby Bartlett Mountain. In-house geologists and mining engineers then inspected the deposits and as a result AMCo signed option agreements for the properties. The owners, one Heckendorf and his associates, retained 25%, and AMCo held 75%. However, following an old Metallgesellschaft/Merton tradition not to burden the company with excessive risk, the investment was syndicated out so that, in the end, AMCo only retained 7.5%, while a number of individuals, chiefly directors and officers of AMCo, acquired the rest. Ludwig Vogelstein, Edward H Clark, the president of Cerro de Pasco Corporation, as well as the British shareholders of AMCo, all declined to participate in the molybdenum venture.

Molybdenum suddenly became scarce because of substantial war demand, and the Bartlett Mountain venture promised to work out. In the summer of 1917, the syndicate decided to spend $300 000 on a mill, and production began in February of 1918. The month before, January 1918, had seen the incorporation of Climax Molybdenum Company. The name came from a small railroad station in the Freemont Pass at the foot of Bartlett Mountain.

But only a few months after operations began, the Climax mill had to close again. Demand for molybdenum as a steel hardener suddenly and dramatically collapsed. Russia, a major buyer, was in the throes of the Bolshevik Revolution and stopped taking deliveries. The war ended and the substantial stocks became immovable.

The mine remained closed for five years, until 1924. In 1925, the US Bureau of Mines and the Association of Automotive Engineers published technical papers praising molybdenum. Another breakthrough came in 1933, when it was shown that molybdenum could replace tungsten.

The 1929–33 depression brought another decline in sales, though only a temporary one. Improved economic conditions and preparation for World War II soon brought a new boom to be again followed by a sharp decline between 1945 and 1950.

The American Metal Company recovered quickly from the 1929–33 depression and soon thereafter became one of the world's largest sellers of non-ferrous metals. It expanded into Africa and Canada.

On 31 December 1957, the Climax Molybdenum Company and the American Metal Company merged to become American Metal Climax, Inc or AMAX. The original capital subscription of $200 000, in 1887 for the formation of AMCo had become assets totalling $250 million. By 1961, the assets had grown to $400 million. Nine years later, in 1970, the assets of AMAX totalled $1.1 billion, to rise to $3.9 billion by 1979, and to $4.9 billion by 1990. Sales in 1990 reached $3.8 billion, with net earnings of $226 million.[30]

The true giant

The latter part of the twentieth century has been a period of rapid changes, all over the world and especially in the United States. And the mining industry has not been spared.

Difficult price and market patterns, competing materials, mergermania, acquisition fever, pollution problems, and changing general economic conditions have brought about substantial changes in the industry. To trace and portray them all would require another book. The following is a summary.

ASARCO is still around, and very strong. Sales in 1990, though at about the same level as in 1989 and 1988, were more than double those of 1986. But the company is still principally a mining and smelting group of non-ferrous minerals and metals. In 1990, ASARCO accounted for 12% of the western world's copper, 14% of the silver, and lead, and 9% of the zinc. And ASARCO still has substantial investments in Mexico and Peru.

Worsening metal prices affected ASARCO's 1991 results; net earnings for the year were a third of those for 1990. The company reached agreement with the government of Peru late in 1990, and pending issues were resolved. Peru agreed to cease discriminating against the company, and a major modernization and investment programme was to be undertaken by Southern Peru Copper Corporation.

Anaconda no longer exists as an independent operating firm. The mine at Butte was shut down. A new company called Montana Resources has partially reopened some old workings, but it is a smaller operation and not comparable to the old Anaconda. Rumour had it that ASARCO had bought a controlling interest in Montana Resources, but this could not be verified.

Anaconda's underground mines lost their competitive edge as early as the 1960s; a gigantic open pit operation seemed the answer. But it was the answer for only a relatively short time. When the 1 mile wide Berkeley Pit closed in 1982, and when the pumps stopped working, a gigantic lake formed, full of acids, metal particles and giant-size pollution problems. The last owner of Anaconda was, and remains, ARCO, Atlantic Richfield, the Los Angeles-based oil conglomerate.

Kennecott was for some 10 years owned by SOHIO – Standard Oil of Ohio – but it was then sold to Rio Tinto Zinc (RTZ). They are the owners of Bingham Canyon, the large mining and smelting complex in Utah, and still one of the largest copper producers in the United States.

Bunker Hill, the large mining and smelting complex in and around Kellogg, Idaho, shut down years ago and is no more.

Newmont Mining spun out its copper operations, which then became Magma Copper Corporation. Magma is new, upcoming, and growing, but 1991 was not a good year for it. Its income was a net loss equal to the net profit of the two previous years. Newmont Mining has become what the company itself calls 'the pre-eminent gold producer in North America' (Annual Report 1990).

AMAX has become 'a major producer of aluminum and manufacturer of aluminum products with substantial interests in domestic (US) energy and gold production'. Its subsidiary, Alumax Inc, was the third largest integrated aluminium company in the United States in 1990, and 80% of the earnings of the corporation came from aluminium. Though still a producer of molybdenum, aluminium is obviously the company's first concern.

Homestake Mining is another of the big mining companies specializing in gold. For it too, 1991 was not a good year, with substantial losses.

Cyprus Minerals Company is the second largest US copper producer, one of the top 20 US coal producers, and the world's leading producer of lithium. Its earnings in 1991 were down from previous years, but it still showed a profit.

And, not to be forgotten among the big copper producers, in the United States and internationally, is the Phelps Dodge Corporation. It is the largest producer in the United States, the second largest in the world, and accounts for fully one-third of all copper produced in the United States.

This brief summary shows how different companies have tried to cope with changing circumstances. AMAX, for example, has become primarily an aluminium company; others have concentrated on gold. A few companies have disappeared. But the gigantic size of the American mining industry is still all there, participating in the growth of the country, delivering the metals vital for growth, and preparing for a new century.

Notes

1. The *New Encyclopaedia Britannica*, 15th edition, 1979, volume 12, page 246.
2. Martinez, *Gold Rushes of North America*, page 31.
3. Corey and Corey, *Exploring the Piedmont of North Carolina*.
4. Martinez, page 33.
5. Van der Zee and Van der Zee, *A Sweet and Alien Land*, page 27.
6. Jackson, *Gold Dust*, page 11.
7. Ibid, pages 12–13.
8. Ibid, page 21.
9. Once more, the reader must be warned that old records were vague, and the figures are at best reasonable estimates.
10. Jackson, page 315. A census taken late in 1850 gave 92 597 people, without counting Indians and obviously only an approximation, what with many of the miners being illiterate and constantly on the move. The 1852 census gave 264 435 people.
11. Ibid, page 325.
12. Lyman, *The Saga of the Comstock Lode*, page x.
13. Ibid, page 5.
14. Ibid, page 8.
15. Davis, *The Guggenheims*, page 47. Mr Davis' book on the Guggenheims is, of course, a classic. It outlines in masterly fashion the involvement of the family in the development of mining in three different countries.
16. Ibid, page 72.
17. Ibid, page 76.
18. Ibid, page 111.
19. Ibid, page 117.
20. Ibid, page 121.
21. Ibid, page 125.
22. Ibid, page 133.
23. Marcosson, *Anaconda*, page 18.
24. Ibid, page 25.
25. Ibid, page 31.
26. Ibid, page 46.

27. Cordero and Tarring, *Babylon to Birmingham*, page 461.
28. Marcosson, page 48.
29. Ibid, page 56.
30. Bernfeld, *A Short History of American Metal Climax, Inc.* and annual reports.

CHAPTER 15

A few loose ends . . .

After reviewing the metal mining industries of the different countries of the Americas throughout the last five centuries, and with some speculation as to what there may have been before the Spaniards and the Portuguese arrived at the end of the fifteenth century, a few loose ends remain.

Several of these 'ends' are fundamental questions; some are controversial issues. A few of the answers are obvious, others are once more controversial. They should be addressed in a forthright and outspoken manner.

Nationalism, chauvinism, internationalism and expropriations

Whose are these lands, the Americas? I, as a European who has lived in the Americas only 35 years, may have no right to call these countries, or any one of them, my own.

But what about the others, those that are here one, or two, or three generations? Would they have a right to invoke, proclaim, claim, and fight for a nation, a nationhood, a nationalism?

Or what about those whose ancestors came 300 or so years ago? Would they be entitled to call these lands their own?

But how did these original settlers acquire land; how did they become owners? Did they take the land? From whom? From the natives? The natives were those erroneously called 'Indians' who inhabited the Americas when the first sailors, adventurers, conquistadors came, some 500 years ago. And would these natives have, or have had, maybe more right to title to these lands than the new arrivals? Why? Because they were here first? First by how long, by 100, or maybe 10 000 years?

'Possession is ownership,' goes the old saying. But how did possession come about? By somebody, at some point in time, 50, or 300, or 10 000 years ago, simply 'taking possession?'

Nationalism, and its somewhat ugly cousin, chauvinism, must thus be considered very subjective, limiting, and not very logical approaches to 'nationhood', and to 'national identity'.

That 'the mines and other natural resources are ours', only makes sense if 'possession is ownership', and to impress the simple minds of the – largely – disenfranchized segments of a given country's population.

This should not be interpreted to mean that 'what should be ours is really everybody's'; such an opposite extreme is probably as wrong as the narrow, limiting nationalism.

Still, there should be food for thought in the calculation that the average price of a basket of metals was, in 1990, about 75% of the average price 10 years earlier, in 1980.[1] Thus, did it make sense to leave the metals in the ground for 10 years, to 'keep such natural resources for future generations'? How many poor starved in those 10 years so that those who proclaimed such theorems could be proven wrong?

A sovereign nation should have and, of course, does have and usually exercises, the right to establish its investment philosophy, whether it is private or state capital that owns industry and develops mines, and whatever controls and taxes the state exercises and levies. And this includes the right of a nation to expropriate assets owned by foreigners or nationals for 'the best of the common good'.

The Spanish crown exercised this right in the case of the Santa Barbara mercury mine in Peru in the sixteenth century, as did the Portuguese crown almost 200 years later in Minas Gerais. And, in more recent times, Mexico, Chile, Peru, Cuba, Bolivia and Brazil have, in different forms, nationalized, expropriated, brought about forced sales and so forth in their mining industries.

The results can at best be called doubtful. Governments and government agencies have again and again revealed themselves to be poor administrators, poor keepers of public trust and public money.

Governments have no business to operate mines or smelters.

And where expropriations involved foreign owners and were not at once followed by 'fair compensation', the country that had expropriated suffered, usually for years, and often even longer, from international ostracism, with foreign capital shying away from new investments, even in other industries.

This even applied – with reason and logic – to a country like Brazil which, with its new constitution of 1988, made foreign capital less than welcome and, as a consequence, saw foreign investments dwindle.

Money, and investments in general, both national and foreign, are volatile; they like to go where they are treated well. And where they are not treated well – whatever the reasons or philosophy behind unfriendly treatment – they evaporate; they run.

Did the discovery and the conquest of the Americas by the Europeans lead to genocide?

The answer to this question is yes, and without any doubt. The native populations of the different regions of the Americas were decimated beyond our comprehension. The historical records are clear; much of the native population simply disappeared.

A few of the early statistics have been discussed in earlier chapters, which looked at what happened in Hispaniola, in Mexico, in Colombia, Peru, Bolivia, and also in Brazil.

It matters little what the original number of inhabitants may have been – 20 million, 30 million or even 100 million. Nobody knows. Nobody counted the natives at the time the first Europeans arrived in the Americas. And the counting that was done later, until well into our century, must also remain suspect, as the criteria used were generally not recorded.

But that vast numbers of natives died should not be doubted. And in some regions as many as 80% or even 90% perished.

But it was not a conscious, wanton murder or extermination comparable to what was done in Germany and Europe under Hitler. The natives of the Americas died first and foremost because the Europeans brought their illnesses against which the early Americans had no antidotes, no acquired immunity. They simply succumbed.

Many also died because of ill-treatment by the Europeans. The conquistadors and their successors considered the natives heathen, little more than animals, and thus not deserving of any consideration.

The hunger – and greed – for gold thus caused the Spaniards to work the natives from dawn to dusk, often without food, and those that dropped out exhausted were forced back to work until they died. In the beginning, the natives were often not allowed to take time off to look after their crops for their own subsistence.

And later, many were forced to work in some mines until they perished, as we know from the Huancavelica mercury mine in Peru.

Life meant little in those times. The ill-treatment meted out to the natives was only a reflection of the general callousness that prevailed in human relations in those days. We only have to read about the general abandon with which the Spaniards killed each other during their long-lasting civil wars in Peru from 1537 to 1554 to realize that a life meant little then – never mind whether that of a native or that of a Spaniard.

Our condemnation of the callousness of the early European visitors (or invaders, or immigrants) should not, however, take us too far. The natives –

or many of the natives – of the Americas were no angels living blissfully in a kind of tropical paradise. The Aztecs were, after all, generous practitioners of human sacrifice, and large pyramids were found, at the time of the conquest by the Spaniards, decorated with tens of thousands of human skulls of the victims of earlier sacrifices. And in Colombia they ate their enemies. One of the chroniclers wrote of what looked like butcher shops, with limbs and bones piled up outside.

The genocide – if we insist on using the term – perpetrated upon the native populations of the Americas by the Europeans after the conquest should thus be looked upon as a fact, a deplorable fact, one to be severely condemned, but still as an act in accordance with the morals and customs of the times.

Natives committed similar exterminations of their neighbours in Mexico and Colombia. Even the Incas transplanted entire nations (or large tribes) from what is Bolivia today to present-day Equador – with much suffering. And in Europe, the Catholic and Protestant armies decimated much of the population of Central Europe during the orgy of cruelty and killing which came to be called the Thirty Years War (1618–48).

Did mining contribute to the development of the Americas?

The answer to this one is decidedly yes!

Greed for wealth led to the first conquests and the subsequent search for treasure. The persistent search for more, and for the sources of the treasures of the pre-Columbians, led more and more to the discovery of new mines.

And then it led to mining. And an outcropping that led to a mine soon also led – if the mine was rich enough – to a settlement, and to a town.

The Americas are thus dotted all over with towns and cities that owe their origin to a mine or a placer, or several placers or several mines right there where the city is today, or nearby. Huancavelica, Cerro de Pasco and La Oroya in Peru, Potosí in Bolivia, and Ouro Preto, São João del Rey, Cuiabá and Diamantina in Brazil are only a few examples, as are Sacramento in California, Butte and Helena in Montana, Trail in British Columbia, as well as Tasco and Guanajuato in Mexico. And there are many more; it is an almost endless list.

The Spaniards and the Portuguese, and later the British, the French and the Dutch, brought settlement, development and progress.

Practically all of the institutions, customs, industry and commerce, as well as the arts, of the Americas of today followed or grew out of the occupation of the Americas by the European immigrants during the last five centuries. They are often tempered, modified, inspired and improved upon by pre-Columbian elements, but they are all essentially European in origin.

It could be questioned just how much good all this did. The development of the Americas has had its downside too: what about the masses of poor, rural and urban, in today's Mexico, Peru, Brazil and the other countries, even in the United States?

In a more narrow sense, mining and the relentless search throughout the centuries for ever more mines, contributed massively to the development of practically all of the countries of the Americas and was instrumental in opening up the interior, or farther reaches of quite a few of the countries.

Thus it was, as we have seen, the gold rush into Minas Gerais that caused the opening up, development and settlement of a vast region in the interior of Brazil. In fact, a good case can probably be made that it was the rush into Minas Gerais, and subsequently into Goiás, Mato Grosso and Mato Grosso do Sul, as well as into Paraná and Santa Catarina, that 'made' Brazil, as these mostly almost empty areas were searched – admittedly only for gold – and then settled, and became known as part of Brazil, from 1700 to 1800. At the end of the period, there was a Brazilian nation.

Something very similar happened in the United States between 1840 and 1880, when the run for gold into California, Nevada, Colorado, and then Idaho, Montana, New Mexico and Utah, made for the *de facto* occupation by Americans, and the incorporation into the Union, of vast lands west, north-west and south-west of the Mississippi river.

Less striking but none the less true are the examples of Bolivia, Peru and Chile.

The northern part of modern Chile, north of the 24th parallel, used to belong to Bolivia, and the northernmost end to Peru. It was quite empty land, but then enterprising Chileans – businessmen, miners, peons – saw that nitrates which were literally lying on the ground, could be picked up and – provided they could be transported to a port and then loaded on to a ship – could bring money.

The Chileans set up an industry in this no man's land over which Bolivia and Peru claimed sovereignty but over which, for lack of manpower and lack of roads or railroads, they could not and did not exercise control. The solution came when Bolivia and Peru finally woke up and tried to control the operations and levy taxes. But Chile invaded to protect its citizens and their industry; Chile won the war and annexed the area.

Yes, mining has contributed mightily to the development of most of the countries of the Americas.

The wealth that came from Nevada and Colorado contributed to the finanicing of the Union during the American Civil War.

And still today, much of the foreign currency income of countries such as Peru, Bolivia, Chile and Brazil comes from the export of mining products.

Still, though the answer to the question is emphatically affirmative, it should be tempered with some further observations.

The towns of Cerro de Pasco, La Oroya and many others, as well as entire countries, like the Peru of 1990, offer vivid testimony that little of the mining wealth taken out of the ground has so far remained to bring wealth and beauty to the surroundings. Most of these industries give and have given work, thus feeding ever larger populations; maybe that is progress and development.

Has foreign capital contributed in meaningful ways to the development of the Americas?

Again, the answer to this question is an emphatic yes.

Domestic capitalists did not want to invest – put risk capital – into the copper mines of Chile. Foreign capital did, and opened up the mines at the beginning of the century.

Foreign capital also invested massively in the mining industry of Mexico, to a point where, for a while, foreigners controlled 97–98% of the industry. Was this bad? No, it was positive! These investments gave jobs to countless Mexicans and built up a great industry. It was only bad when the industry was left, for a while, unchecked and uncontrolled.

And in Peru, it was again foreign capital that opened up much of the central sierra region – Cerro de Pasco and La Oroya.

In Brazil, it was a mixed story of foreign interest and somewhat later national capital. But Belgo-Mineira was the first large scale steel works, and even the huge Itabira iron ore deposits, around the Rio Doce valley, were in the beginning in foreign hands, until CVRD was set up and took on an ever larger role. Finally, even the Carajás complex was discovered by foreigners, and held by them until they sold to Brazilian interests.

A different course was taken in the United States, where mainly domestic capitalists developed the mining industry. In Canada, much American capital contributed, though Canadian capitalists and money did not shy away from investing.

Only Bolivia was different; there genuine native capitalists developed two of the dominant mining groups, the Aramayos and Patiño.

The role of foreign capital in the development of the mining industry – or any industry – of a country should not be looked upon with antagonism. Capital, of any origin, is a potential asset anywhere. If it is treated well, it comes in, contributes to development, pays taxes, creates jobs and becomes an integral part of the fabric of a nation.

Is the role of mining declining as the twentieth century draws to a close?

The answer here is also yes.

Houses, refrigerators, bridges, cars, and many other commonplace goods in our daily lives used to consist, 50, 20, 10 and even 5 years ago, of much more metal – steel, copper, lead, zinc, aluminium – than they do today. The trend seems irreversible. Yes, as humanity grows in numbers, overall production of goods increases, and though each car today uses less metal, more cars in total make up for much of the reduction in the use of metals. Still, metals are bound to play a lesser role as time goes on.

Ten – or was it 15? – years ago, doomsday was predicted, and the forecast was that humanity would run out of a number of metals before the end of the century. It does not look that way any more . . .

And in some metals, like iron ore, prices have declined over the last 20 or 30 years.

Conclusion

This is a book of history and of stories.

Can there be an objective, impartial, view of history, a balanced review of what happened in the past?

The answer must be 'no'.

Our view of history, or the view of a particular author, depends not only on the selective bias of the individual, but also on how selective the historical memory has been.

The further back we go, the less we know, and the more accidental are the details that have survived the forgetfulness of time.

Time buries; time destroys – especially that which happened very long ago, before people wrote down what they knew and wanted to preserve records for posterity.

Our knowledge of what the aborigines of the Americas knew of mining and metallurgy prior to the arrival of the Spaniards, the Portuguese, the British, the French and the Dutch, is very fragmentrary. We depend on the 'stories' the conquistadors and their chroniclers recorded – and they recorded generally what they wanted to write down, and how they understood what they were told.

And we depend on what is found today, and how we interpret these finds.

There is no absolute truth in history. Much of it appears to be just a collection of stories.

Individuals shaped history, those that stood or stand out, whose names we remember. Others, many others, unknown people, also shaped history. We know they existed, though we do not know their names. And accidents, events, sudden turns of fate, illnesses, storms, cold spells, floods – they all made history. And some are remembered, in the form of stories, but most are forgotten.

Mining in the Americas is an attempt to capture a few of these stories; history is but a series of stories, like pearls on a string.

Note

1. Here are the composite prices of six very basic metals, as taken from the statistical yearbooks of Metallgesellschaft; the prices are said to be the yearly averages:

		1980	1990
Aluminium	¢ lb USA 99.5% ingots	69.6	74.0
Lead	¢ lb USA common grade, NY	42.5	46.0
Copper	¢ lb USA elec. Wbars, US prod.	101.4	121.8
Zinc	¢ lb USA PWZ, East St Louis	37.4	74.6
Tin	¢ lb USA Straits tin	846.0	386.3
Nickel	¢ lb USA 999 fine	341.5	406.8
Average		239.7	184.9

Averaging tin with copper and mixing aluminium with lead may not really make much sense. The question may also be asked why include nickel, and not silver, iron ore or any other metal or metals? It seems arbitrary. Still, any other list, even if much more complete and with a weighted average, could probably be equally problematic. I prefer to let the above list stand, as a mere illustration.

Bibliography

Latin American authors with double last names are listed, when from Spanish-speaking countries, under the first of such double names, for example Mario Castro Arenas appears as Castro Arenas, Mario. In the case of Brazilian authors, the last family name appears first, i.e. Francisco Magalhães Gomes is listed as Gomes, Francisco Magalhães. I have tried to be consistent in both cases, listing authors by their father's family name.

Magazine and newspaper articles are listed separately.

Sources identified with an asterik (*) are primary.

Abecia, Valentin. *Historia de Chuquisaca*. Sucre, Bolivia: Editorial Charcas, 1939.

Abreu, Capistano de. *Caminhos Antigos e Povoamento do Brasil*. São Paulo: Editora Itatiaia Ltda, 1989.

*Agricola, Georg. *Vom Berg und Huettenwesen (De Re metallica, libri XII)*. Muenchen: Deutscher Taschenbuchverlag, 1977.

Aguirre Lavayén, Joaquín. *Guano Maldito*. La Paz: Editorial Los Amigos del Libro, 1976.

Alexander, William and Arthur Street. *Metals in the Service of Man*. Harmondsworth, Middlesex: Pelican Books, 1977.

Alvear Urrutia, Jorge. *Chile. Nuestro Cobre*. Santiago de Chile: Editorial Lastra SA, 1974, and London: Granada Publishing Ltd, 1969.

Antonetti, Guy. *A Economia Medieval*. São Paulo: Atlas, 1977.

Antze, Gustavo. *Trabajos en Metal en el Norte del Peru*. Lima: Universidad Nacional Mayor de San Marcos (UNMSM), 1965.

Arciniegas, Germán. *Geschichte und Kultur Lateinamerikas*. Muenchen, Nymphenburger Verlagshandlung GmbH, 1966.

Biografia del Caribe. Buenos Aires: Editorial Sudamericana, 1959.
Arellano Moreno, Antonio. *Origens de la Economia Venezolana*. Caracas: Ediciones de la Biblioteca, Universidad Central de Venezuela, 1973.
Atkin, Ronald. *Revolution: Mexico 1910–20*. London: Granada Publishing Ltd, 1969.
Ballantine Cobb, Gwendolyn. *Potosí y Huancavelica*. La Paz: Biblioteca 'Bamin', Academia Boliviana de la Historia, 1977.
Bandeira, Manuel. *Guia de Ouro Preto*. Rio de Janeiro: Editora Tecnoprint SA, 1985.
Barton, Robert. *A Short History of Bolivia*. La Paz: Editorial Amigos del Libro, 1968.
Bassols Batalla, Angel. *Recursos Naturales de Mexico*. Mexico, DF: Editorial Nuestro Tiempo SA, 1981.
Baudin, Louis. *Der sozialistische Staat der Inkas*. Hamburg: Rowohlt, 1956.
Daily Life in Peru under the last Incas. London: G Allen & Unwin Ltd, 1961.
Bauer, P T. *Equality, the Third World and Economic Delusion*. Cambridge, Mass.: Harvard University Press, 1981.
Belli, L Prospero. *La Civilizacion Nazca*. Lima: Talleres Graficos P L Villaneuva, SA, 1960.
Benavides, M Julio. *Historia de la Moneda de Bolivia*. La Paz: Empresa Editora Universo, 1972.
Bendek, Frank. *Evolucion de la Economia Dominicana desde 1966 y sus possibilidades futuras*. Santo Domingo: Editora Corripio, C por A, 1988.
*Benzoni, M Jeronimo. *La Historia de Nuevo Mundo*. Lima: Universidad Nacional Mayor de San Marcos (UNMSM), no date.
Bernfeld, Seymour S (in collaboration with Harold K Hochschild) *A Short History of American Metal Climax, Inc*: New York, about 1961 or 1962.
Bernstein, Marvin D. *The Mexican Mining Industry, 1890–1950*. New York: University of New York, 1965.
Bingham, Hiram. *Lost City of the Incas*. New York: Duell, Sloan & Pearce, 1948.
Blakemore, Kenneth. *The Book of Gold*. New York: Stein & Day, 1971.
Bonavia, Duccio *et al. Pueblos y Cultura de la Sierra Central del Peru*. Lima: Cerro de Pasco Corporation, 1972.
Bonilla, Heraclio. *Guano y la Burguesia en el Perú*. Lima: Instituto de Estudios Peruanos (IEP), 1974.
Booz-Allen & Hamilton do Brazil, Consultores Ltda. *Minerals & Metals Outlook Brazil 1977*. Rio de Janeiro: Photocopy, 1977.
Borda, Fals. *As revolucões inacabadas na America Latina*. São Paulo: Global Editora, 1979.
Bosson, Rex and Bension Varon. *The Mining Industry and the Developing Countries*. New York: Oxford University Press, 1977. Published for the World Bank.
Bourne, Edward Gaylord. *Spain in America, 1450–1580*. New York: Barnes & Noble Inc, 1962.
Braudel, Fernando. *The Mediterranean and the Mediterranean World in the Age of Philip II*. New York: Harper & Row, 1972.
Bray, Warwick. *Gold of El Dorado*. New York: Harry N Abrams, Inc, 1979.
Burns, E Bradford. *A History of Brazil*. New York, Columbia University Press, 1970.
Buse, H. *Huaras – Chavin*. Lima: Juan Mejia Baca, 1957.

Machu Picchu. Lima: Libreria Studium, 1978.
Bushnell, G H S. *Peru, Ancient Peoples and Places*. Lima: ABC Bookstore SA and London: Thames & Hudson, 1956.
Busto, D Jose Antonio del. *Dos Personajes de la Conquista del Peru*. Lima: Editorial Universitario SA, 1969.
Peru Preincaico. Lima: Editorial Universitario SA, no date.
*Cabildo de la Ciudad de San Juan de la Frontera de Huamanga. *Libro de 1539–1547*. Lima: Casa de la Cultura del Peru, 1966.
Cañete, Pedro Vincente. *Historia Fisica y Politica de la Provincia de Potosí*. La Paz: Fundacion Universitaria 'Simon I Patiño', 1952.
Capriles Villazon, Orlando. *Historia de la Mineria Boliviana*. La Paz: Biblioteca 'Bamin', 1977.
Carnac, Pierre. *A Atlantida de Cristovão Colombo*. Rio de Janeiro: Diefel, 1978.
*Casas, Bartolomé de las. *Historia de las Indias*. Mexico: Fundo de Cultura Economica, 1951.
**O Paraiso Destruido*. Porto Alegre: L & PM Editores Ltda, 1984.
Cassá, Roberto. *Historia Social e Economica de la Republica Dominicana*. Santo Domino, RD: Editora Alfa e Omega, 1987.
Castro Arenas, Mario. *La Rebelión de Juan Santos*. Lima: Milla Batres, 1973.
Ceram, CW. *Der Erste Amerikaner*. Hamburg: Rowohlt, 1972.
Cesar, Guilhermino. *Minas Gerais, Terra e Povo*. Porto Alegre: Editora Globo, 1970.
Cespedes, Augusto. *Relatos de la Guerra del Chaco*. Lima: Populibros Peruanos, no date, but about 1970.
Cheetham, Sir Nicolas. *A Short History of Mexico*. New York: Thomas Crowell Company, 1970.
*Chronicle of Colonial Lima. *The Diary of Joseph and Francisco Mugaburu 1640–1697*. Norman, Oklahoma: University of Oklahoma Press, 1975.
*Chronicle. *La Aventura del Amazonas*. Carvajal, G de Almesto Y, P de Rojas, Alonso de Rafael Diaz, Editor. Madrid: *Historia* 16, 1986. 1986.
*Cieza de Leon, Pedro de. *Auf den Koenigsstrassen der Incas*. Stuttgart: Steingrueben Verlag, 1971.
**La Cronica del Peru (Parte Primera)*. Mexico: Editorial Nueva España, 1950(?). (Original published Seville: Martin de Montesdoca, 1553.)
*Colombo, Cristovão (Columbus). *Diarios de la Descoberta da America*. Porto Alegre: L & PM Editores Ltda, 1964.
Consejo de Recursos Minerales. *Anuario Estadistico de la Minería Mexicana, 1979*. Mexico, DF: Consejo de Recursos Minerales, 1980.
Sumario Estadistico de la Minería Mexicana 1983–1987. Mexico, DF: Consejo de Recursos Minerales, 1988.
Cordero, H G and L H Tarring. *Babylon to Birmingham*. London: Quin Press Ltd, no date.
Corey, Jane and Corey, Cindy. *Exploring the Piedmont of North Carolina*. Chapel Hill, NC: The Provincial Press, undated, about 1989/90.
Corporacion del Cobre (Chile). *El Cobre Chileno*. Santiago: Editorial Universitaria, 1975.

*Cortés, Hernando. *Five Letters to the Emperor*. New York: W W Norton & Co Inc, 1962.
Costa Villavicencio, Lázaro. *Historia Cronologica del Peru, Años 1000–1919*. Lima: Imp y Lit Salesiana, no date.
Crespo, Alfonso. *Los Aramayos de Chichas: Tres Generaciones de Mineros Bolivianos*. Barcelona: Editorial Blume, 1981.
*Cronica Inedita de Comienzos del Siglo XVII. *Decripcion del Virreinato del Peru*. Rosario, Argentina: Universidad Nacional del Litoral, 1958.
Dale, Jackson Donald. *Gold Dust*. New York: Alfred A Knopf, 1980.
Dalton, Leonard V. *Venezuela*. London: T Fisher Unwin Ltd, 1916.
Dauhajre hijo, Andrés. *Sabado Economico*. Santo Domingo, RD: Fundacion Economia y Desarollo Inc, 1990.
Davis, John H. *The Guggenheims: An American Epic*. New York: William Morrow & Co Inc, 1978.
Denker, Cornelia, *Gold in Brasilien*. Staden-Jahrbuch 1980, Sao Paulo, Brazil.
Descola, Jean. *Daily Life in Colonial Peru, 1710–1820*. London: George Allen & Unwin Ltd, 1968.
*Días del Castillo, Bernal. *Historial Verdadera de la Conquista de la Nueva España*. Madrid: Coleccion Austral, Espasa-Calpe SA, 1985.
Disselhoff, Hans Dietrich. *Vida en el Antiguo Peru*. Muenchen: Librerias ABC SA, 1970.
Eschwege, W L von. *Pluto Brasiliensis*. Belo Horizonte: Livraria Itatiaia Editora Ltda, 1979.
Fawcett, Brian. *Railways of the Andes*. London: George Allen & Unwin Ltd, 1963.
Flores Polo, Pedro. *Geografia Economica del Peru*. Lima: Corp Editora Continental SA, 1973.
Franco, Francisco de Assis Carvalho. *Dicionario de bandeirantes e sertanistas do Brasil*. São Paulo: Itatiaia Ltda/USP, 1989.
Frank, Andre Gunder. *Capitalism and Underdevelopment in Latin America*. New York: Monthly Review Press, 1969.
Fundação Memorial da América Latina. *O Indio/Ontem Hoje/Amanhã: Dossiê de l Ciclo*. São Paulo: Eda Tassara, Org Edusp, 1991.
Galeano, Eduardo. *As Veias Abertas da America Latina*. São Paulo: Paz e Terra, 1979.
*Garcilaso de la Vega, 'Inca' (real name: Gomez Suarez de Figueroa). *Obras Completas*. Madrid: Ediciones Atlas, 1965.
Geddes, Charles F. *Patiño, the Tin King*. London: Robert Hale, 1972.
Giardini, Cesare. *The Life and Times of Columbus*. Feltham, Middlesex: The Hamlyn Group Ltd, 1967.
Goff, Jacques Le. *Mercaderes e Banqueiros da Idade Média*. Lisbon: Gradiva, 1982(?).
Gomes, Francisco Magalhães. *Historia da Siderurgia no Brasil*. Belo Horizonte: Editora Itatiaia Ltda, 1983.
Green, Timothy. *The World of Diamonds*. New York, William Morrow & Company, Inc, 1981.
*Guaman Poma de Ayala, Felipe. *La Nueva Cronica y Buen Gobierno*. Lima: Editorial Cultura, 1956.

Guasch, Juan. *Bolivar El Eterno Traicionado*. Barcelona: Editorial Mateu, 1961.
*Guillén, Edmundo. *Version Inca de la Conquista*. Lima: Editorial Milla Batres, 1974.
Habich, Eduardo de. *Los Fenicios en la Historia del Peru*. Lima: Ediciones Eduardo de Habich, 1970.
Hagemann, Helmut. *Hohe Schornsteine am Amazonas*. Freigurg: Dreisam-Verlag, 1985.
Hagen, Victor von. *Heeresstrassen des Sonnengottes*. Vienna: Ullstein & Co, 1957.
Auf der Suche nach dem Goldenen Mann. Hamburg: Rowohlt, 1977.
Realm of the Incas. New York: New American Library, 1957.
World of the Maya. New York: New American Library, 1960.
The Aztec: Man and Tribe. New York: New American Library, 1958.
Der Ruf der Neuen Welt: Deutsche bauen Amerika. Muenchen: Droemer Knaur, 1970.
Manuelas Jahreszeiten der Liebe. Manuela Saenz und Simon Bolivar. Hamburg: Paul Szolnay Verlag, 1957.
Halperin Dongi, Tulio. *Historia Contemporanea de America Latina*. Madrid: Alianza Editorial, 1988.
Hanke, Lewis. *La 'Historia de la Villa Imperial de Potosí', por Bartolomé Arzáns de Orsúa y Vela*. Lima: Universidad Nacional Mayor de San Marcos (UNMSM), no date.
Hanson, Earl Parker. *South from the Spanish Main*. New York: Delacorte Press, 1967.
Hemming, John. *The Conquest of the Incas*. London: Macmillan London Ltd, 1970.
Henseling, Karl Otto. *Bronze, Eisen, Stahl: Bedeutung der Metalle in der Geschichte*. Hamburg: Rowohlt, 1984.
Hoffmann, Arnold. *Free Gold: The story of Canadian Mining*. New York: Associated Book Service, 1947.
Homet, Marcel F. *Chanchan: La Misteriosa*. Barcelona: Ediciones Martinez Roca SA, 1977.
Horkheimer, Dr Hans, *La Cultura Mochica*. Lima: Peruano Suiza SA, 1961.
La Cultura Inca. Lima: Cia de Seguros Peruano Suiza, 1965.
Houben, HH. *Christoph Columbus*. Berlin: Volksverband der Buecherfreunde, Wegweiser Verlag, GmbH, 1932.
Huber, Siegfried. *Pizarro und seine Brueder*. Olten: Walter Verlag, 1962.
Humboldt, Alexander von. *Vom Orinoko zum Amazonas*. Wiesbaden: F A Brockhaus, 1964.
Tagebuch vom Orinoko. Berlin: Verlag der Nation, 1966.
Huntington, Samuel P. *Political Order in Changing Societies*. New Haven: Yale University Press, 1968.
Instituto Brasileiro de Siderurgica (IBS). *Brazilian Iron and Steel Industry*. Rio de Janeiro: IBS, 1984.
Instituto Cultural de Providencia. *El Cobre Chileno en el Tiempo*. Santiago de Chile and Barcelona: Empresa Industrial Grafica, no date.
Instituto de Ingenieros de Minas del Peru. *XI Convencion: Algunos Aspectos de Tecnica Minera del Peru*. Lima, 1969.
Inter American Development Bank. *Annual Report 1978*. Washington, DC: IADB, 1979.
International Monetary Fund (IMF). *International Financial Statistics*. Washington, DC: IMF, July 1990.

Izard, Miguel. *Tierra Firma: História de Venezuela y Colombia*. Madrid: Alianza Editorial SA, 1987.
Jackson, Donald Dale. *Gold Dust*. New York: Alfred A Knopf, 1980.
Jeier, Thomas. *Der grosse Goldrausch von Alaska*. Freiburg: Herder KG, 1972.
*Jerez, Francisco de. *Conquista del Peru y Provincia del Cuzco*. Madrid: Editorial Nueva Espana, SA, 1950(?) (Original published in Madrid, 1534.)
Jones, Lawrence F and Lonn, George. *Historical Highlights of Canadian Mining*. Toronto: Pitt Publishing Co Ltd, 1973.
*Juan, Jorge y Antonio de Ulloa. *Noticias Secretas de America*. Madrid: Ediciones Turner, 1982. (Original published in 1826.)
Kandell, Jonathan. *La Capital: The Biography of Mexico City*. New York: Random House, 1988.
Kaufmann Doig, Federico. *Arqueologia Peruana*. Lima: Peisa, 1971.
El Peru Antiguo. Lima: Sociedad Academica de Estudios, 1963.
Knauth, Percy. *The Metalsmiths*. New York: Time Life Books, 1974.
Kolata, Alan L. 'Tiwanaku and its hinterland'. Boston, *Archaeology Magazine*, Jan/Feb 1987.
Konetzke, Richard (ed). *Fischer Weltgeschichte*, Band 22, *Sued- und Mittelamerika I*, Frankfurt: Fischer Buecherei, 1965.
Kotschko, Ricardo. *Serra Pelada*. São Paulo: Editora Brasiliense, 1984.
Kutscher, Gerd. *Chimu*. Berlin: Verlag Gebrueder Mann, 1950.
Langdon, Gee. *Don Roberto's Daughter*. London: Chatto & Windus, 1968.
Lanning, Edward P. *Peru before the Incas*. Englewood Cliffs, NJ: Prentice-Hall, Inc, 1967.
Latif, Miran de Barros. *As Minas Gerais*. Rio de Janeiro: Livraria Agir Editora, 1960.
*Leme, Pedro Taques de Almeida Paes. *Noticias das Minas de São Paulo e dos sertões da mesma capitania*. São Paulo: Editora Itatiaia Ltda/USP, 1980.
León-Portilla, Miguel. *La Minería en Mexico*. Mexico: Universidad Nacional Autónoma de Mexico, 1978.
Lewin, Boleslao. *La Inquisicion en Hispano America*. Buenos Aires: Paidos, 1967.
Los Judios bajo la Inquisicion en Hispanoamerica. Buenos Aires: Editorial Dedalo, 1960.
Lindquist, Sven. *Suedamerika: Der gepluenderte Kontinent*. Muenchen: Humboldt Taschenbuchverlag, 1972.
Linke, Lilo. *Ecuador*. London: Oxford University Press, 1960.
Lohmann Villena, Guillermo. *Las Minas de Huancavelica en los Siglos XVI y XVII*. Seville: Escuela de Estudios Hispano-Americanos de Sevilla, 1949.
Lonn, George. *Historical Highlights of Canadian Mining*. Toronto: Pitt Publishing Company Ltd, 1973.
Lopes, Margaret and Silvia Figueroa. *O Conocimento Geologico na America Latina*. Campinas: IG UNICAMP, 1990.
Lopez, Luiz Roberto. *Historia do Brasil Colonial*. Porto Alegre: Mercado Aberto Editora e Propaganda Ltda, 1983.
*López de Gómara, Francisco. *Historia General de las Indias*. Barcelona: Editorial Iberia SA, 1966. (Original published, 1552.)

Lopez Rosado, Diego G. *Problemas Economicos de Mexico*. Mexico, DF: Universidad Nacional Autonoma de Mesico, 1979.

Lothrup, S K. *El Tesoro del Inca*. Lima: Universidad Nacional Mayor de San Marcos (UNMSM), 1938.

Lumbreras, Luis Guillermo. *Los Templos de Chavin*. Lima: Corporacion Peruana del Santa, 1970.

Los Origenes de la Civilizacion en el Peru. Lima: Editorial Milla Batres, 1977.

Lyman, George. *The Saga of the Comstock Lode: Boomdays in Virginia City, 1859–1865*. New York: Ballantine Books, 1971.

Lynch, John. *The Spanish-American Revolutions, 1808–1826*. New York: W W Norton & Co Inc, 1973.

Machado, Iran F. *Recursos Minerales, Politica e Sociedade*. São Paulo: Editora Edgard Bluecher Ltda, 1989.

Machado Filho, Aires da Mata. *Arraial do Tijuco: Cidade Diamantina*. São Paulo: Editoria Itatiaia Ltda/USP, 1980.

Madariaga, Salvador de. *Christoph Columbus*. Stuttgart: Europaeischer Buchclub, 1940.

Die Erben der Conquistadoren. Stuttgart: Deutsche Verlagsanstalt, 1964.

Malpica, Carlos. *Los Duenos del Peru*. Lima: Peisa, 1970.

Manso, Gilberto Costa. *O caso da Indústria do Aluminio: Un Estudo de Organização Industriál*. Brasília: Universidade de Brasília, Departamento de Economia, 1985.

Marcosson, Isaac F. *Anaconda*. New York: Dodd, Mead & Co, 1957.

*Markham, Sir Clements R. *Reports on the Discovery of Peru: Francisco de Xeres, Miguel de Astete, Hernando Pizarro, Pedro Sancho*. London: Hakluyt Society, 1872.

The Incas of Peru. Lima: ABC, 1977.

Marmontel, M. *Les Incas*. Paris, 1810.

Marques, Maria Luiza de Aguiar. *A Industria de Aluminio no Brasil*. Rio de Janeiro: UFRJ, 1983.

Marshall, Edison. *Cortez and Marina*. New York: Popular Library, 1963.

Martin, Luis. *The Kingdom of the Sun*. New York: Charles Scribner's Sons, 1974.

Martinez, Lionel. *Gold Rushes of North America*. Secaucus, NY: The Wellfleet Press, 1990.

Mason, F van Wyck. *Golden Admiral*. Garden City, NY: Doubleday & Co Inc, 1953.

Mason, J Alden. *The Ancient Civilizations of Peru*. Harmondsworth, Middlesex: Penguin Books Ltd, 1969.

Maza Zavala, D F, y otros. *Venezuela, Crecimiento sin Desarollo*. Caracas: Editorial Nuestro Tiempo SA, Universidad Central de Venezuela, 1982.

Venezuela, una Economia Dependiente. Caracas: Instituto Universitario de Tecnologia Antonio José de Sucre, 1985.

Mazur, Suzan. 'Visions of the Alto Magdalena', Boston, *Archaeology Magazine*, Nov/Dec 1989, page 30.

Meggers, Betty J. *Ecuador*. London: Thames & Hudson, 1966.

Meira, Silvia. *Fronteira Setentrional: 3 Seculos de Lutas no Amapá*. São Paulo: Editora Itatiaia Ltda, 1989.

Mello e Souza, Laura de. *Desclassificados do Ouro: A Pobreza Mineira do Seculo XVIII*. Rio de Janeiro: Ediçoes Graal, 1986.

Menzel, Dorothy. *La Cultura Huari*. Lima: Cia de Seguros Peruano-Suiza SA, 1968.
Metallgesellschaft Aktiengesellschaft. *Metallstatistik 1967–1990*, various yearbooks. Frankfurt: Metallgesellschaft.
Milling-Stanley, George with Jessica Jacks. *Gold 1989*. London: Consolidated Gold Fields PLC, 1989.
Miró Quesada, S Aurelio. *Garcilaso y otros Estudios Garcilacistas*. Madrid: Ediciones Cultura Hispanica, 1971.
Mirsky, Jeanette. *The Gentle Conquistadors*. New York: Random House, 1969.
Moraes, Walfrido. *Jagunços e Heróis* (4th edition). Salvador/Bahia: Empresa Grafica da Bahia/IPAC, 1991.
Morisse, Dr Lucien. *Excursion a El Dorado – El Callao*. Caracas: Cochano Films, for Corporacion Venezolana de Guayana, 1985.
Mosley, Michael E (ed). *Chanchan, Andean Desert City*. Albuquerque, NM: University of New Mexico Press, 1982.
Moya Pons, Frank. *Después de Colón: Trabajo, Sociedad y Politica en la Economia del Oro*. Madrid: Alianza Editorial SA, 1987.
Historia Colonial de Santo Domingo. Santiago, RD: Universidad Catolica Madre e Maestra, 1977.
Mujica Gallo, Miguel. *Catalogue, Museum Gold of Peru*. Lima: Miguel Mujica Gallo Foundation, 1970.
Norddeutsche Affinerie (ed). *Copper*. Hamburg: Norddeutsche Affinerie, 1966.
Northern Miner Press Limited. *Mining Explained*. Toronto: Northern Miner, 1968.
Noya Pinto, Virgilio. *O Ouro Brasileiro e o Comercio Anglo-Portugues*. São Paulo: Companhia Editora Nacional, 1979.
Nunes Leal, Paulo. *O Outro Brazo da Cruz*. Rio de Janeiro: Cia Brasileira de Artes Graficos, 1986.
*Nuñez, Estuardo (ed). *4 Viajeros Alemanes al Peru*. W Bayer, K Scherzer, F Gerstaecker, H Zoeller. Lima: Universidad Nacional Mayor de San Marcos (UNMSM), no date.
El Peru en la Obra de Alejandro de Humboldt. Lima: Librearia Studium, 1971.
Núñez Cabeza de Vaca, Alvar. *Naufragios*. Madrid: Alianza Editorial, 1985.
Nussbaum, Bruce. *The World After Oil*. New York: Simon & Schuster, 1983.
Ogger, Guenter. *Kauf Dir einen Kaiser: Die Geschichte der Fugger*. Muenchen-Zuerich: Droemersche Verlagsanstalt Th Knaur Nachf, 1978.
Olivia de Coll, Josefina. *A Resistencia Indigena*. São Paulo: L & PM Editores Ltda, 1974.
Palma, Ricardo. *Tradiciones Peruanas*. Madrid: Aguilar SS de Ediciones, 1968.
*Parker, Franklin. *Travels in Central America, 1821–1840*. Gainesville, Florida: University of Florida Press, 1970.
Pease, G Y Franklin. *Del Tiwantinsuyu a la Historia del Peru*. Lima: Instituto de Estudios Peruanos, 1978.
Peña, Sergio de la. *La Formacion del Capitalismo en Mexico*. Mexico: Siglo Veintiuno Editores SA, 1979.
Petersen, Dr Georg. *Mineria y Metalurgia en el Antiguo Peru*. Lima: Museo Nacional de Antropologia y Arqueologia, 1970.

Peterson, Mendel. *The Funnel of Gold.* Boston–Toronto: Little, Brown & Company, 1975.
Piekalkiewicz, Janusz. *Da liegt Gold.* Muenchen: Suedwestverlag, 1971.
Pike, Frederick B. *The Modern History of Peru.* London: Weidenfeld & Nicolson, 1967.
Pinillos, R A. *Chanchan.* Trujillo: Ediciones Oro Chimu, 1972.
*Pizarro und andere Conquistadoren, 1526–1712. *Die Eroberung von Peru,* Celso Garcia, Gaspar de Carvajal, Samuel Fritz. Tuebingen: Horst Erdmann Verlag, 1973.
*Plinius Secundus d A C. *Naturkunde: Buch XXXIII Metallurgie.* Muenchen: Artemis Verlag, 1984.
Pons Muzzo, Gustavo. *Historia del Peru.* Lima: Editorial Universo SA, 1961.
Popenoe, Dorothy H. *The Story of Antigua Guatemala.* Dalton, MA: Studley Press, 1973.
Poppino, Rollie E. *Brazil, the Land and the People.* New York: Oxford University Press, 1968.
*Porras Barranechea, Raul. *Las Relaciones Primitivas de la Conquista del Peru.* Lima: Instituto Raul Porras Barranechea, 1967.
Prain, Sir Ronald. *Copper.* London: Mining Journal Books Ltd, 1975.
Prescott, William H. *History of the Conquest of Mexico* and *History of the Conquest of Peru.* New York: The Modern Library, about 1950.
Prieto, Carlos. *A Mineraçao e o Novo Mundo.* São Paulo: Editora Cultrix, 1976.
Puhle, Hans-Juergen (ed). *Lateinamerika: Historische Realitaet und Dependencia Theorien.* Hamburg: Hoffmann & Lampe Verlag, 1977.
Purser, W F C. *Metal-Mining in Peru, Past and Present.* New York/London, Prager Publisher Inc, 1971.
Querejazu Calvo, Roberto. *Llallagua: Historia de una Montaña.* La Paz: Editorial Los Amigos del Libro, 1977.
Rama, Angel. *Der lange Kampf Lateinamerikas: Texte und Dokumente von José Marti bis Salvador Allende.* Frankfurt: Surkamp, 1982.
Ravines, Rogger (ed). *100 Años de Arqueologia en el Peru.* Lima: IEP y Petroperu, 1970.
Reiche, Maria. *Mystery in the Desert.* Stuttgart: Nazca Peru SA, 1968.
Riverend, Julio Le. *Historia Economica de Cuba.* La Havanna: Editorial de Ciencias Sociales, 1985.
Robertson, William Spence. *Rise of the Spanish American Republics (As told in the lives of their liberators).* New York: Collier Books, 1961.
Rodriguez Hoyle, Daniel (ed). *Peru Minero 1967.* Lima: Instituto de Ingenieros de Minas del Peru, 1968.
La Mineria Metalica Peruana, 1971. Lima: Editorial Litografico La Confianza SA, 1972.
Rodriguez Martinez, Lorenzo. *Panoramica de la Isla Hispaniola, Siglos XI–XVI.* Sto Dominogo, Republica Dominicana: Taller, 1989.
Roel, Virgilio. *Historia Social y Economica de la Colonia.* Lima: Editorial Grafica Labor, 1970.
Los Libertadores. Lima: Editorial Grafica Labor, 1971.
Romero, Emilio. *Historia Economica del Peru.* Lima: Editorial Universo SA, 1949.

Ruiz Estrada, Arturo. *Arqueologia de la Ciudad de Huancavelica*. Lima: Servicios de Artes Graficas SA, 1977.

Rutledge, Ian with Phil Wright. 'Colombia plans gold revival'. London, *Financial Times*, 4 September 1987.

*Salvador, Frei Vincente do. *Historia do Brasil 1500–1627*. São Paulo: Editora Itatiaia/ Edusp, 1982.

Salvador, José Gonçalves. *Os Cristãos-Novos: Povoamento e Conquista do Solo Brasileiro 1530–1680*. São Paulo: Livraria Pioneira/Edusp/USP, 1976.

Os Magnates do Trafico Negreiro. São Paulo: Livraria Pioneira/Edusp, 1981.

Samamé Boggio, Mario. *El Peru Minero*. Lima: Editora Peru, 1979.

Mineria Peruana. Lima: Editorial Litografico La Confianza SA, 1972.

Sames, Carl Wolfgang. *Die Zukunft der Metalle*. Frankfurt: Suhrkamp, 1974.

Sanchez, Luis Alberto. *La Perricholi*. Lima: Universidad Nacional Mayor de San Marcos (UNMSM), 1963.

Sanchez-Albornoz, Nicolás. *Indios y Tributos en el Alto Peru*. Lima: IEP Ediciones, 1978.

Santos, Breno Augusto dos. *Amazonia: Potencial Mineral e Perspectivas de Desenvolvimento*. São Paulo: T A Queiroz, Editor Ltda, 1983.

'*Carajás – historia e perspectivas*'. A report dated Brasilia, 27 March 1984.

*Sarmiento de Gamboa, Pedro. *La 'Historia Indica'*. Lima: Universidad Nacional Mayor de San Marcos (UNMSM), 1964.

Sauer, Jules Roger, *Brazil, Paradise of Gemstones*. Rio de Janeiro: Grafica Riex Editora SA, 1982.

Savoy, Gene. *Vilcabamba, Last City of the Incas*. London: Robert Hale, 1970.

*Schmidel, Ulrico. *Relatos de la Conquista del Rio de la Plata y Paraguay, 1534–1554*. Madrid: Alianza Editorial SA, 1986. (Original published in 1602.)

Schurz, William Lytle. *Latin America: A Descriptive Survey*. New York: E P Dutton & Co Inc, 1942.

Semo, Enrique. *Historia del Capitalismo en Mexico, los Origines 1521–1763*. Mexico: Ediciones Era SA, 1973.

Sharp, William Frederick. *Slavery on the Spanish Frontier: The Colombian Chocó, 1680–1810*. Norman, Oklahoma: University of Oklahoma Press, 1976.

Silva, Helio S. *Tendencia y Caracteristicas do Comercio Exterior no Seculo XIX*. Rio de Janeiro: Revista de Historia da Economia Brasileira, June 1953.

Silva Bruno, Ernani. *Historia do Brasil*. São Paulo: Editora Cultrix, 1967.

Simonsen, Roberto C. *Historia Economica do Brasil 1500–1820*. São Paulo: Companhia Editora Nacional, 1978.

Sociedad Nacional de Mineria. *Memorias 1967–1990*. Lima.

Strauss, Simon O. *Trouble in the Third Kingdom*. London: Mining Journal Books Ltd, 1986.

Tamayo Herrera, José. *Historia Social del Cuzco Republicano*. Lima: Industrialgrafica SA, 1978.

Tannenbaum, Frank. *Ten Keys to Latin America*. New York: Vintage Books, 1966.

*Tauro, Alberto (ed). *Viajeros en el Peru Republicano*. Samuel Haigh, Carl August Gosselman, José Victorino Lastarria, C Skogman, Geo W Carleton. Lima: Uni-

versidad Nacional Mayor de San Marcos (UNMSM), no date.
Torres Saldamando, Enrique. *Apuntes Historicos sobre las Encomiendas en el Peru*. Lima: Universidad Nacional Mayor de San Marcos (UNMSM), about 1953.
Trimborn, Herrmann. *El Delito en las Altas Culturas de America*. Lima: Universidad Nacional Mayor de San Marcos (UNMSM), no date. (Original published in 1936.)
*Tschudi, J J von. *Peru: Reiseskizzen aus den Jahren 1838–1842*. Graz, Austria: Akademische Druck & Verlagsanstalt, 1846 (Nachdruck).
Twinam, Ann. *Miners, Merchants and Farmers in Colonial Colombia*. Austin, Texas: University of Texas Press, 1982.
Uhle, Max. *Estudios sobre Historia Incaica*. Lima: Universidad Nacional Mayor de San Marcos (UNMSM), 1969.
*Ulloa, Antonio de y Jorge Juan. *Noticias Secretas de America*. Madrid: Ediciones Turner, 1982. (Original published in 1826.)
United Nations. *Economic Survey of Latin America, 1967*. New York: United Nations Publications, 1969.
Valcarcel, Luis E. *Historia del Peru Antiguo*. Lima: Editorial Universitaria, 1964.
Vargas Ugarte, SJ Ruben. *Historia General del Peru*. Lima: Editor Carlos Milla Batres, 1971.
Vasconceles, Agripa. *Chico Rey*. Belo Horizonte: Editora Itatiaia Ltda, 1966.
Vasconcelos, Diogo de. *Historia Media de Minas Gerais*. Belo Horizonte: Editora Itatiaia Ltda, 1974.
Vega, J J. *Incas, Dioses y Conquistadores*. Lima: Fondo de Cultura Popular, no date.
Manco Inca, el gran Rebelde. Lima: Populibros Peruanos, no date.
José Gabriel Tupac Amaru. Lima: EUNE, 1969.
Vilar, Pierre. *Ouro e Moeda na Historia (1450–1920)*. Rio de Janeiro: Editora Paz e Terra, 1981.
Vives, J Vicens. *Historia de España y America, social y economica*, volume V. Barcelona: Libros Vicens-Bolsillo, Editorial Vicens Vives, 1982.
Wang, Chun T and Stephen W H Yih. *Tungsten, Metallurgy, Properties and Applications*. New York: Plenum Press, 1979.
Womack Jr, John. *Zapata and the Mexican Revolution*. New York: Alfred A Knopf, 1969.
World Bank. *Informe sobre el Desarollo Mundial 1984*. Washington DC: Oxford University Press, 1984.
World Development Report 1987. New York: Oxford University Press, 1987.
Yates, Robert G, Dean F Kent and Jaime Fernandez Concha. *Geology of the Huancavelica Quicksilver District*. Washington DC: US Government Printing Office, 1951.
Yepes del Castillo, Ernesto. *Peru 1820–1920: Un siglo de desarollo capitalista*. Lima: IEP, 1972.
Young Jr, Otis E. *Western Mining*. Norman, Oklahoma: University of Oklahoma Press, 1970.
*Yupanqui, Titu Cusi. *Relacion de la Conquista del Peru*. Lima: Biblioteca Universitaria, 1973. (Original published in 1916.)
*Zarate, Agustin de. *Historia del Decubrimiento y Conquista de la Provincia del Peru*. Mexico: Editorial Nueva Espana SA, 1950(?).

Zee, Henry van der and Barbara van der Zee. *A Sweet and Alien Land: The Story of Dutch New York*. New York: The Viking Press, 1978.

Magazine and newspaper articles

Archaeology

'Columbia's lurid past', by Susan Mazur, Nov/Dec 1989.
'Bolivia's hinterland', by Alan L Kolata, Jan/Feb 1987.

Brasil-Mineral

'As maiores empresas do setor mineral', April 1987.
'Ouro – retomando projetos', March 1991.
'O ouro já no brilha tanto: Ou brilha?', January 1985.
'Os 150 anos da Mina de Morro Velho', June 1984.
'O peso do garimpo', January 1984.

Financial Times

'Colombia plans gold revival', by Ian Rutledge, 4 September 1987.
'Brazil's low cost tin policy', by Patrick Knight, 5 August 1987.

Folha de São Paulo

'Un conselho do barão', by José Sarney (former president of Brazil), 26 July 1991, page 2.

Metal Bulletin

Articles by Diana Kinch, Latin American editor.
'Big rise in Brazil aluminium exports', 2 July 1990.
'CVRD sees alumina as priority', 9 July 1990.
'Bugbear of the world tin market', 20 August 1990.
'Energy costs rein in primary aluminium growth', 20 August 1990.
'Iron ore stands at production peak', 20 August 1990.
'Big drop possible in Brazilian tin output', 28 January 1991.
'Bom Futuro strategy', 4 March 1991.
'Brazil gold mine output to match *garimpos*', 23 May 1991.
'Brazil gold output continues falling', May 1991.
'Gold garimpo era over – DNPM', 30 May 1991.
'Codemin to install nickel sinter plant', 10 June 1991.
'AMSA sees gold mine closure', 20 June 1991.
'Morro Velho continues to invest heavily', 20 June 1991.

Metal Bulletin Monthly

Articles by Diana Kinch, Latin American editor.

'Brazil gets to grips with *garimpos*', April 1991.
'Iron ore – Brazil adjusts to no-growth scenario', April 1991.

Metal Bulletin – Futures and Options World

'Gold trading takes off in Brazil', January 1989.

Mineração Metalurgica

'Os 150 anos da Mina de Morro Velho', May 1984.

National Geographic Magazine

'Moche grave', October 1988.
'Track of the Manila galleons', September 1990.
'The timeless mystique of emeralds', July 1990.
'New Moche tomb', June 1990.

Index